Microbial Biotechnology Approaches to Monuments of Cultural Heritage

Ajar Nath Yadav • Ali Asghar Rastegari •
Vijai Kumar Gupta • Neelam Yadav
Editors

Microbial Biotechnology Approaches to Monuments of Cultural Heritage

Springer

Editors
Ajar Nath Yadav
Department of Biotechnology
Eternal University
Sirmour, Himachal Pradesh, India

Ali Asghar Rastegari
Department of Molecular & Cell Biochemistry
Falavarjan Branch, Islamic Azad University
Isfahan, Iran

Vijai Kumar Gupta
Department of Chemistry & Biotechnology
Tallinn University of Technology
Tallinn, Estonia

Neelam Yadav
Gopi Nath P.G. College
Veer Bahadur Singh Purvanchal University
Ghazipur, Uttar Pradesh, India

ISBN 978-981-15-3400-3 ISBN 978-981-15-3401-0 (eBook)
https://doi.org/10.1007/978-981-15-3401-0

This Springer imprint is published by the registered company Springer Nature Singapore Pte Ltd.
The registered company address is: 152 Beach Road, #21-01/04 Gateway East, Singapore 189721, Singapore

Preface

The countries' cultural legacy is one of the world's most diverse. It is a beacon that draws millions every year to our convents and monuments, to our museums and libraries, as well as to our concert halls and festivals. It is also a very dynamic trigger of economic activities and jobs. Among the different scientific branches, microbial biotechnology allows for an innovative and precise approach to the complexity of the problems that the restorer has to face with his own daily work. This present book, *Microbial Biotechnology Approaches to Monuments of Cultural Heritage*, is a very timely publication, providing state-of-the-art information in the area of microbial biotechnology focusing on microbial biodiversity and their biotechnological application in monuments of cultural heritage. The book volume comprises 10 chapters. In Chap. 1, Kumar et al. describe the microbial community present on the reverse side of a deteriorated canvas. In Chap. 2, Biswajit Batabyal as lead author highlights the microbial biocleaning technologies for cultural heritage. In Chap. 3, Upadhyay et al. describe the role of bacterial communities in the prevention of microbial growth on cultural heritage. In Chap. 4, Vara et al. highlight the opportunities and challenges of entomogenous fungi and the conservation of the cultural heritage. In Chap. 5, Parulekar-Berde et al. describe the biodiversity of microorganisms and their enzymes as biorestoration agents. In Chap. 6, Parulekar-Berde et al. deal with the bioremediation of cultural heritage for the removal of organic substances. Chandra et al. highlight the role of microorganisms in the removal of sulfates on artistic stoneworks in Chap. 7. In Chap. 8, Amrita Kumari Panda and his colleagues describe in detail the microbiological tools for cultural heritage conservation. In Chap. 9, Kumar et al. highlight the microbial biotechnology techniques for restoration and conservation. In Chap. 10, Enshasy et al. explain about the novel approach in the restoration of construction materials for culturable heritage.

This book covers biodiversity of microbial communities from diverse cultural heritage, microbial biotechnological techniques for biocleaning for cultural heritage, role of bacterial fungal communities for conservation of the cultural heritage, and microbial enzymes and their potential applications as biorestoration agents. This book will be immensely useful to biological sciences, especially to microbiologists, microbial biotechnologists, biochemists, researchers, and scientists of microbial biotechnology. We are honored that the leading scientists who have extensive, in-depth experience and expertise in microbial biotechnology took the time and effort to develop these outstanding chapters. Each chapter is written by internationally recognized researchers/scientists, providing an up-to-date and detailed account of our knowledge of the microbial biotechnology approaches to monuments of cultural heritage.

Baru Sahib, Himachal Pradesh, India — Ajar Nath Yadav
Isfahan, Iran — Ali Asghar Rastegari
Tallinn, Estonia — Vijai Kumar Gupta
Ghazipur, Uttar Pradesh, India — Neelam Yadav

Acknowledgments

All authors are sincerely acknowledged for contributing up-to-date information on the microbial biotechnology approaches to monuments of cultural heritage. The editors are thankful to all authors for their valuable contributions.

All editors would like to thank their families who were very patient and supportive during this journey. They sincerely thank the whole Springer team who was directly or indirectly involved in the compilation of this book and Ms. Aakanksha Tyagi, Mr. Ashok Kumar, and Ms. Vaishnavi Venkatesh for their valuable suggestions and encouragement throughout the project.

The editor, Dr. Ajar Nath Yadav, is grateful to his PhD research scholars, Tanvir Kaur, Rubee Devi, Divjot Kour, Geetika Guleria, Chandranandi Negi, and Kusam Lata Rana, and colleagues for their support, love, and motivation in all his efforts during this project.

The editors are also grateful to all the reviewers for their timely support and advice which improved the quality of the book. They are very sure that this book will be of great interest to the scientists, graduates, undergraduates, and postdocs who are investigating on microbial biotechnology approaches to monuments of cultural heritage.

Contents

Editors and Contributors

About the Editors

Ajar Nath Yadav is an assistant professor (Senior Scale) in the Department of Biotechnology, Eternal University, Baru Sahib, Himachal Pradesh, India. He has 10 years of teaching and research experiences. He obtained his doctorate degree in Microbial Biotechnology in 2016, jointly from Indian Agricultural Research Institute, New Delhi and Birla Institute of Technology, Mesra, Ranchi, India. Dr. Yadav has 174 publications with h-index 37, i10-index 76, and 3143 citations. Dr. Yadav has got 12 Best Paper Presentation Awards, and "Outstanding Teacher Award"-2018. He has the lifetime membership of Association of microbiologist in India, Indian Science Congress Council and National Academy of Sciences, India.

Ali Asghar Rastegari is currently working as an Assistant Professor in the Faculty of Biological Science, Department of Molecular and Cell Biochemistry, Falavarjan Branch, Islamic Azad University, Isfahan, I.R. Iran. Dr. Rastegari has 13 years of teaching and research experiences. Dr. Rastegari gained a Ph.D. in Molecular Biophysics, the University of Science and Research, Tehran Branch, IranIn his credit 18 research publications and published 12 abstracts in different conference. He has a lifetime membership of Iranian Society for Trace Elements Research (ISTER), The Biochemical Society of I.R.IRAN, and Member of Society for Bioinformatics in Northern Europe (SocBiN).

Vijai Kumar Gupta from ERA Chair of Green Chemistry, Department of Chemistry and Biotechnology, School of Science, Tallinn University of Technology, Tallinn, Estonia, is one of the leading experts in the area of microbial biology and biotechnology. He is the member of International Sub-commission on Trichoderma and Hypocrea, Austria; International Society for Fungal Conservation, UK; and Secretary of European Mycological Association. Dr. Gupta is the Fellow of Prestigious- The Linnaean Society, London, UK; Fellow of Indian Mycological Association; and Fellow of Mycological Society of India. He has been honored with several awards in his career including Indian Young Scientist Award for his advanced research achievements in the field of fungal biology and biotechnology. He is the editor of leading scientific journals of high repute and having many publications in his hands with h-index 38. He is editor of many books published by international publishers.

Neelam Yadav is senior researcher, currently working on microbial diversity from diverse sources and their biotechnological applications agriculture and allied sectors. She obtained her post graduation degree from Veer Bahadur Singh Purvanchal University, Uttar Pradesh, India. She has research interest in the area of probiotics microbes, and beneficial microbiomes from diverse sources. In her credit 59 publications in different reputed international, national journals and publishers. She is Editor/reviewer of different international and national journals. She has the lifetime membership of Association of microbiologist in India, Indian Science Congress Council, India and National Academy of Sciences, India.

Contributors

Jameel Ahmad Department of Zoology, Gandhi Faizam College, Mahatama Jyotiba Phule Rohilkhand University, Bareilly, Uttar Pradesh, India

Biswajit Batabyal Serum Analysis Center Pvt. Ltd., Kolkata, West Bengal, India

Vikrant B. Berde Department of Zoology, Arts, Commerce and Science College, Lanja, Maharashtra, India

Satpal Singh Bisht Department of Zoology, Kumaun University, Nainital, Uttarakhand, India

Prem Chandra Department of Environmental Microbiology, Babasaheb Bhimrao Ambedkar University (A Central University), Lucknow, Uttar Pradesh, India

P. Veera Bramha Chari Department of Biotechnology, Krishna University, Machilipatnam, Andhra Pradesh, India

Jimmy Ngow Chee Ghong Huashi Malaysia Sdn. Bhd, Kuala Lumpur, Malaysia

Vandna Chhabra Department of Agronomy, School of Agriculture, Lovely Professional University, Jalandhar, India

Daniel Joe Dailin Institute of Bioproduct Development (IBD), Universiti Teknologi Malaysia (UTM), Johor Bahru, Malaysia

School of Chemical Engineering and Energy, Faculty of Engineering, Universiti Teknologi Malaysia (UTM), Johor Bahru, Malaysia

Swathi Dash Department of Environmental Science, GITAM Institute of Science, GITAM (Deemed to be University), Visakhapatnam, Andhra Pradesh, India

Hesham El Enshasy Genetic Engineering and Biotechnology Research Institute, City of Scientific Research and Technology Applications (CSAT), New Burg Al Arab Alexandria, Egypt

Institute of Bioproduct Development (IBD), Universiti Teknologi Malaysia (UTM), Johor Bahru, Malaysia

School of Chemical Engineering and Energy, Faculty of Engineering, Universiti Teknologi Malaysia (UTM), Johor Bahru, Malaysia

Enespa Department of Plant Pathology, School of Agriculture, Sri Mahesh Prasad Degree College, University of Lucknow, Lucknow, Uttar Pradesh, India

Jennifer Eyahmalay Institute of Bioproduct Development (IBD), Universiti Teknologi Malaysia (UTM), Johor Bahru, Malaysia

School of Chemical Engineering and Energy, Faculty of Engineering, Universiti Teknologi Malaysia (UTM), Johor Bahru, Malaysia

Sachin S. Ghoble Department of Zoology, Arts, Commerce and Science College, Lanja, Maharashtra, India

Neha Kapoor Hindu College, University of Delhi, Delhi, India

Manoj Kumar Karnena Department of Environmental Science, GITAM Institute of Science, GITAM (Deemed to be University), Visakhapatnam, Andhra Pradesh, India

Ho Chin Keat Institute of Bioproduct Development (IBD), Universiti Teknologi Malaysia (UTM), Johor Bahru, Malaysia

Lamha Kumar Hindu College, University of Delhi, Delhi, India

Rajesh Kumar Department of Zoology, Gandhi Faizam College, Mahatama Jyotiba Phule Rohilkhand University, Bareilly, Uttar Pradesh, India

Sushil Kumar Department of Botany, Shaheed Mangal Pandey Govt Girls PG College, Meerut, Uttar Pradesh, India

Upendra Kumar Department of Molecular Biology, Biotechnology and Bioinformatics, College of Basic Sciences and Humanities, CCS Haryana Agricultural University, Hisar, India

Rajesh Lalloo Council for Scientific and Industrial Research (CSIR), Bioprocess Development Group, Pretoria, South Africa

Roslinda Abd Malek Institute of Bioproduct Development (IBD), Universiti Teknologi Malaysia (UTM), Johor Bahru, Malaysia

School of Chemical Engineering and Energy, Faculty of Engineering, Universiti Teknologi Malaysia (UTM), Johor Bahru, Malaysia

Rojita Mishra Department of Botany, Polasara Science College, Polasara, Ganjam, Odisha, India

Nurul Zahidah Nordin Institute of Bioproduct Development (IBD), Universiti Teknologi Malaysia (UTM), Johor Bahru, Malaysia

School of Chemical Engineering and Energy, Faculty of Engineering, Universiti Teknologi Malaysia (UTM), Johor Bahru, Malaysia

Amrita Kumari Panda Department of Biotechnology, Sant Gahira Guru University, Ambikapur, Chhattisgarh, India

Chanda Parulekar-Berde Department of Microbiology, Gogate Jogalekar College, Ratnagiri, Maharashtra, India

Priyanka Department of Botany, Government Girls Degree College, Kharkhauda, Meerut, Uttar Pradesh, India

Santosh Ramchuran Council for Scientific and Industrial Research (CSIR), Bioprocess Development Group, Pretoria, South Africa

Veshara Malapermal Ramdas School of Life Sciences, University of KwaZulu-Natal, Durban, South Africa

Prachiti P. Rawool Department of Microbiology, Gogate Jogalekar College, Ratnagiri, Maharashtra, India

Sagar P. Salvi Department of Microbiology, Gogate Jogalekar College, Ratnagiri, Maharashtra, India

R. Sanjana Department of Environmental Science, GITAM Institute of Science, GITAM (Deemed to be University), Visakhapatnam, Andhra Pradesh, India

Jatinder Singh Department of Horticulture, School of Agriculture, Lovely Professional University, Jalandhar, India

Rishikesh R. Surve Department of Chemistry, Arts, Commerce and Science College, Lanja, Maharashtra, India

Archana Tiwari Amity University, Noida, Uttar Pradesh, India

Hina Upadhyay Department of Agronomy, School of Agriculture, Lovely Professional University, Jalandhar, India

Saritha Vara Department of Environmental Science, GITAM Institute of Science, GITAM (Deemed to be University), Visakhapatnam, Andhra Pradesh, India

[illegible]

Microbial Community Present on the Reverse Side of a Deteriorated Canvas

1

Sushil Kumar, Priyanka, and Upendra Kumar

Abstract

Cultural heritage makes the foundation of any nation. Culture includes melodies, music, dance, theater, people conventions, performing expressions, ceremonies and customs, compositions, and artworks. Painting is a significant sort of art. Painting is the practice of applying paint, pigment, color, or other medium to a solid surface. Paintings are made on different kinds of supports like paper, pastel, parchment, canvas, walls, and roofs of monuments. Paintings on canvas need canvas as support, a preparatory layer of vegetal or animal glues, a paint layer that includes the pigment and its binder (most commonly used is linseed oil), and finally a layer of varnish to protect the pigment layer. Different layers are composed of many organic and inorganic compounds. Biodeterioration is the alteration of organic and inorganic materials induced by the metabolic activity and growth of microorganisms. Paintings on canvas are prone to biodeterioration. Biodeteriogens secrete an array of aggressive metabolic products, viz., organic and inorganic acids, as well as hydrolytic enzymes that degrade the substratum. Fungi and bacteria are the major causes of biodeterioration of canvas paintings. Some of the cellulolytic fungal genera that degrade the canvas support are reported to involve one or more species: *Aspergillus*, *Fusarium*, *Myrothecium*, *Memnoniella*, *Neurospora*, *Alternaria*, *Penicillium*, *Scopulariopsis*,

S. Kumar
Department of Botany, Shaheed Mangal Pandey Govt Girls PG College, Meerut, Uttar Pradesh, India

Priyanka
Department of Botany, Government Girls Degree College, Kharkhauda, Meerut, Uttar Pradesh, India

U. Kumar (✉)
Department of Molecular Biology, Biotechnology and Bioinformatics, College of Basic Sciences and Humanities, CCS Haryana Agricultural University, Hisar, Haryana, India

A. N. Yadav et al. (eds.), *Microbial Biotechnology Approaches to Monuments of Cultural Heritage*, https://doi.org/10.1007/978-981-15-3401-0_1

Stachybotrys, *Chaetomium*, and *Stemphylium*. On the other hand, bacteria include genera *Bacillus* and *Staphylococcus*. Genera like *Staphylococcus*, *Acinetobacter*, *Agrococcus*, *Janibacter*, *Rhodococcus*, and *Stenotrophomonas* are cultured and isolated on artificial media from the deteriorated canvas.

Keywords

Biodeterioration · Biodeteriogens · Canvas painting · Artwork · Cultural heritage

1.1 Introduction

Development of a nation depends greatly on its culture. The culture includes shared mindset, customs, qualities, objectives, and practices. Culture and innovativeness show themselves in practically all monetary, social, and different practices. A nation as diverse as India is symbolized by the majority of its way of life. India has one of the world's biggest accumulations of melodies, music, move, theater, people conventions, performing expressions, ceremonies and customs, works of art, and compositions that are known as the "Elusive Cultural Heritage" of mankind. Art has genuine incentive to the social orders that make it, and this implies it has different capacities. Frequently, art is utilized for the reason for narrating, regardless of whether that implies composing, painting, singing, moving and so on. Besides that, occasionally art plays a significant role in the scholarly culture of a society. Workmanship comprises of an assortment of fine arts, including painting, model, ceramics, and material expressions, for example, woven silk. Painting is a significant sort of art.

Painting is the practice of applying paint, pigment, color, or other medium to a solid surface (https://en.wikipedia.org/wiki/Painting). Among the most ancient paintings are at the Grotte Chauvet in France; it is believed that these are about 32,000 years old. The lowermost layer of materials, ochre, used is estimated about 60,000 years. Archeologists have additionally discovered a section of shake painting saved in a limestone shake cover in the Kimberley district of North-Western Australia that is 40,000 years old (aboriginalartonline.com). Biodeterioration is the alteration of organic and inorganic materials induced by the metabolic activity and growth of microorganisms (Di Carlo et al. 2017).

It is now an established fact that microorganisms with certain environmental factors degrade culturally significant artifacts. Development of microbial communities in artworks is largely affected by the constitution of the support, i.e., canvas, wood, paper, etc., and quantity of humidity and nutrients (Saiz-Jimenez 1993). Micro-fungi generally require relative humidity higher than 65% and a range of temperature between 20 and 300°C (Garg et al. 1995). In paintings the preparatory layer is most vulnerable due to its hygroscopic nature. By and large this layer is swollen, appearing as crystals with no stability and no adherence to the canvas.

1.2 Biodeterioration of Paintings

The support materials for paintings may be canvas, wood, paper, or parchment. A preparatory layer on support is prepared as per the kind of painting. This preparatory layer in canvas paintings is prepared with salts of calcium such as lime (calcium oxide and calcium hydroxide) or gypsum (hydrated calcium sulfate) in combination with glue of either animal or plant origin. The paint layers are then applied on the smooth side of preparatory layer. The paints may be of different kinds like oil paints, distemper, or water colors according to the objective. The layer of colors contains mixture of pigments and binders of oil or distemper, i.e., egg or glue. At last a thin translucent layer of varnish for defensive purposes is applied on the obverse face. Finally, the paint on the obverse side is usually covered with a protective layer of thin translucent varnish.

Biodeterioration of artworks may involve either partially or all the parts of the painting. Heterotrophic biodeteriogens use organic substances for their nutrition; there are a number of organic compounds in different layers of paintings, i.e., support, preparatory, or paint layer with binders. These organic substances provide a good habitat and nutrition to the heterotrophic biodeteriogens, mostly bacteria and fungi. These biodeteriogens accumulate and multiply to degrade the artworks under favorable conditions which are easily available in museum rooms and old churches and in collections with no control on temperature and humidity. Paintings provide a variety of organic compounds for accumulation and multiplication of all the species of micro-fungi. Micro-fungi consist of a greater degree of tolerance to the fluctuations in the surrounding environment. Additionally, they use water condensed on the surface of the paintings. The water content is not sufficient for bacterial growth. In comparison to micro-fungi, bacteria are unable to utilize condensed water (Dhawan and Agrawal 1986).

Microbial deterioration of paintings on canvas generally begins from the support, i.e., the reverse side. This is due to the organic nature of the canvas and the glue sizing. The glue sizing is naturally multiplying the susceptibility of canvas to microbial attack. The biodeteriogens start degradation of support canvas and finally reaches the back side of the paint layer. Hydrolysis of cellulose and organic components of canvas and preparatory layer by micro-fungi and/or bacteria causes the detachment of paint layer from the support. This also causes cracks and discoloration of paint layer (Ravikumar et al. 2012).

The microorganisms involved in biodegradation of organic materials are also participating in biodeterioration of painting on canvas due to the similarity in compounds of vegetable and animal origin present on the reverse side. The preparatory layer due to its organic and inorganic components increases the susceptibility manifolds (Dhawan and Agrawal 1986; Makies 1981). By comparison, the paint layer is more resistant to microbial attack than the support. This is due to the nature of the pigments. Casein and egg distempers are among the most susceptible to microbial attacks followed by emulsion distemper. Linseed oil is least susceptible in this context. The resistance to biodegradation of pigment layer can be increased by the addition of heavy metals like lead, zinc, and chromium in the pigments. Water colors are highly prone to biodeterioration because there is a little amount of organic binders in them (Dhawan and Agrawal 1986).

Both reverse and obverse faces of a canvas painting are degraded by variety of biodeteriogens. On the obverse side, deterioration magnitude is a function of the paint or pigment utilized and the way of its application; meanwhile on the back side, i.e., the reverse face, it results due to the composition of supporting material, viz., canvas or wood. The process of biodeterioration on oil paintings generally starts on the reverse side because the support materials and the sizing glue provide plenty of organic and inorganic substances for the growth of microorganisms (Tiano 2002). At the same time, the organic matter on the obverse side of a painting is liable to be attacked by special kind of microorganisms. This may be due to occasional contaminants, like transient airborne microorganisms. Bacteria and fungi produce spores which may remain attached to the obverse side for a long time and constitute the sub-aerial community that can accumulate on this side of painting for a long time in the form of spores. These microorganisms then grow and degrade the substances; as a result the paint layer get detached from the support. This occurs under highly humid and warm conditions (Ciferri 1999; Schabereiter-Gurtner et al. 2001).

Biodeteriogens secrete an array of aggressive metabolic products, viz., organic and inorganic acids, as well as hydrolytic enzymes that degrade the substratum (Ciferri 1999). The main hydrolytic enzymes involved in deterioration of paintings are lipases and esterase. Different kinds of lipases and estrases are secreted by an array of microorganisms. Lipases hydrolyse triglycerides by breaking ester bonds, while esterases degrade substrates soluble in water (Soliman et al. 2007), endo-N-acetyl-glucosaminidases hydrolyze mucopeptides or chitin (Karamanos 1997).

The knowledge of microorganisms associated with a deteriorated artifact plays a significant role in developing the effective strategies for their restoration and conservation (Schabereiter-Gurtner et al. 2001; Capodicasa et al. 2010; Suihko et al. 2007). Biodeterioration of mural and frescoes is studied widely. The findings described different microbial communities associated with the obverse side of paintings (Ciferri 1999; Schabereiter-Gurtner et al. 2001; Pinar et al. 2001; Gonzalez and Saiz-Jimenez 2005).

Microbial communities on canvas paintings are less explored, though (Capodicasa et al. 2010). To study microbial communities associated with the biodeterioration of paintings, samples were collected from the areas where visible signs of deterioration appeared. There may be alterations in the pigment colors, presence of spots, or differences in the structure of the obverse layer. Standard samples were collected from the parts where no damage is observed. Indoor air quality was analyzed by collecting and investigating the air samples. Culture-dependent and culture-independent techniques are two major strategies employed to study the microbial communities responsible for degradation (Laiz et al. 2003).

Biodeteriogens are responsible for degradation of artworks and monuments (Peterson and Klocke 2012). The organic and inorganic components in the artwork make an optimal medium for microbial colonization; as a result aesthetic and structural symptoms appear on the artwork (Ranalli et al. 2009). To design optimal strategies for conservation and restoration of artworks, it is necessary to understand the microbial diversity and composition of materials used in the artwork because chemoorganotrophic, chemolithotrophic, and photolithotrophic microbes depend on the chemical nature of their substratum (Capodicasa et al. 2010). A number of studies on damaged stone monuments reported microbial colonization (Videla et al. 2000; Mihajlovski et al. 2017) and also on mural paintings (Guglielminetti et al. 1994; Gorbushina et al. 2004; Rosado et al. 2017) and frescoes (Sampa and Luppi Mosca 1989; Radaelli et al. 2004), but very few studies have reported the microbial communities present on easel paintings, namely, on canvas (Lopez-Miras et al. 2013a, b; Capodicasa et al. 2010; Pavic et al. 2015; Salvador et al. 2017) or wood panel (Rubio and Bolivar 1997; Pellerito et al. 2016).

Easel paintings or paintings on canvas generally provide the wide range of microhabitat and organic and inorganic nutrients which are suitable to microbial growth. Canvas paintings are composed of thin overlapped layers of pigments with binders spread on canvas support. The pictorial layer remains in between the layer of varnish and the preparatory layer (Taft and Mayer 2000; Stulik and Paint 2000). Several organic (pigment, binder, and varnish) and inorganic (pigment) are present in these layers (Stulik and Paint 2000; Leonardi 2005; Matteini and Mazzeo 2009). A wide range of microbe can grow in the support material like canvas, pigment, oil, animal glue, or the compounds used for adhering the pictorial film to the preparatory layer, e.g., oil and animal glue. Biodeterioration is also accelerated by specific environmental conditions, i.e., warm and humid conditions, because such conditions are suitable for growth and stimulate the production of aggressive metabolism (Capodicasa et al. 2010; Pinna and Salvadori 2008, Sassu et al. 2017). The objective of this review is to investigate work done up to date on less explored and very important part of conservation and restoration of culturally important paintings, i.e., the microbial communities present on the reverse face of an oil or easel painting on canvas.

This is commonly observed that the paintings are not placed in such environment that may prevent the development of biodeteriogens. This is the need of the hour to explore the microbes involved in deterioration of canvas paintings. By using innovative strategies, we can find the changes and treatments needed in order to manage the diverse materials employed in artwork correctly (Poyatos et al. 2017).

1.3 Microbial Communities in Biodeterioration

There is a variety of biodeteriogens on different artworks, monuments, and paintings of cultural importance. Biodeteriogens may belong to different categories such as bacteria, cyanobacteria, fungi, algae, lichens, mosses, and higher plants. These all may cause severe damage to culturally important artworks. Growth and multiplication of microbial communities on artifacts depends on habitat and physico-ecological conditions. Different microbial communities inhabit various ecological niche and promote deterioration of organic and inorganic components of the materials of cultural heritage in their own way. The deterioration of oil paintings generally starts from the reverse side and not on the obverse side due to the resistance of materials used in painted layer and varnish. Ecesis of microbes is favored by environmental conditions in combination with organic and inorganic substances that provide energy and nutrients (Urzì and Krumbein 1994).

1.3.1 Algae and Cyanobacteria

Photosynthetic microorganisms, viz., algae, cyanobacteria, and photosynthetic bacteria, are reported on the surface artworks which are exposed to sunlight or artificial light and also dampness, warmth, and inorganic nutrients. These are the pioneers on artifacts of cultural importance. These organisms cause aesthetic damage to the artwork and create the conditions suitable for the successful establishment, growth, and multiplication of other microbes (Urzì and Krumbein 1994). Algae, growing on monuments and murals, secrete a small quantity of organic acids, sugars, and proteins and cause minute deterioration to the artwork. These products favor the growth of bacteria with a potential to cause serious deterioration to the substratum (Di Carlo et al. 2017). Algae and cyanobacteria cause nutrient enrichment of the substratum by accumulating organic carbon and entrapping dust and soil particles (Hirsch et al. 1995; Souza-Egipsy et al. 2004). Blue-green algae may survive under unfavorable high temperature and low light intensity conditions (Lamenti et al. 2003). Cyanobacteria are capable to produce pigments that protect them from damaging effects of UV rays and adapted to survive under drying rehydration cycles (Albertano et al. 2003; Hoffmann 2002; Roldan et al. 2004). This category of biodeteriogens is rarely reported from the deteriorated reverse side of the canvas paintings. This may be due to the absence of light as well as other favorable conditions on the reverse side of the paintings.

1.3.2 Fungi as Biodeteriogens of Canvas Paintings

Fungi are a group of heterotrophic osmotrophs that belong to the kingdom *Mycota*. The paintings are usually displayed on the walls of museums and other places. The dampness and warmth are important factors for the establishment of fungal biodeteriogens on the reverse face of the canvas paintings. The spore and conidia from different species of bacteria remain suspended in the atmosphere due to their size, shape, and water. These spores and conidia when come in contact with the support material of a painting under high humidity and warm temperature got germinated via a germ tube. Such conditions generally prevail in the tropical regions.

Germination of spores leads the formation of hyphae and then mycelium in or on the substratum. The growth of mycelium exerts the mechanical pressure and distortion in the canvas. The mycelium that remains on the surface of the canvas causes aesthetic damage. The aggressive metabolic products degrade the organic components in the canvas as well as in the preparatory layer of vegetable or animal glue.

The fungi synthesize and release the enzymes in the medium. These extracellular enzymes hydrolyze the complex organic components of the medium and absorb the nutrients by diffusion (osmotrophs). In the Earth's carbon cycle, decomposition of lignin and cellulose is a major event. It is necessary for steady state of an ecosystem and the Earth as a whole. Fungi are considered the primary decomposers (Alexander 1981). Fungal enzymes are economically very significant to mankind. Maximum destruction of food occurs due to molds. Fungi destroy cellulosic as well as non-cellulosic substances that may be a part of a canvas painting (Onions et al. 1981; Rose 1981).

In biodeterioration of painted materials, fungi in association with certain insects are also often reported to cause damage of the support and rarely cause harm to the obverse painted side; wooden stretchers, preparatory layer of glue, and the canvas are in priority to be deteriorated. Some important fungal genera secrete cellulolytic, lipolytic, and proteolytic enzymes that hydrolyze the different components of the paintings on canvas. Some cellulolytic fungal genera that include one or more species are *Alternaria*, *Aspergillus*, *Fusarium*, *Memnoniella*, *Myrothecium*, *Neurospora*, *Penicillium*, *Scopulariopsis*, *Stemphylium*, *Stachybotrys*, and *Chaetomium*. The coloration of the fungal mycelia also studied, e.g., in *Dematiaceae*, has been reported to cause aesthetic damage (Caneva et al. 1991).

Micro-fungi those frequently deteriorate the painted layer belong to the species affiliated to the genera *Trichoderma*, *Penicillium*, *Aspergillus*, *Phoma*, etc. These preferentially degrade distemper and oil binders. Oil binders are degraded by *Aureobasidium*, *Geotrichum* develop on casein binders, and *Mucor* and *Rhizopus* grow on and decompose the glue in the preparatory layer of canvas paintings (Gallo 1985). The aesthetic, mechanical, and biochemical damages are due to the growth and secretions of micro-fungi. The mycelia of different fungal species grow over the surface of the paint layer and hide the color patterns. Growth of such fungi also occurs under the paint layer and produces fruiting bodies in turn destroying the preparatory layer and the canvas on the reverse side of the paintings. Such changes in the organization of paint and bolster layers cause friability and loss of the paint layer. Presence of pigments in the fruiting bodies and colored metabolites made the permanent colored spots, and hydrolyzing substances cause irreversible changes in the structure of the paintings.

1.3.3 Bacteria as Biodeteriogens of Canvas Paintings

Various reports prove that the bacterial communities establish prior to the fungal ones. (Santos et al. 2009). Bacteria, the prokaryotic organisms, remain freely or in the form of colonies. Based on their shapes, bacteria can be categorized under different groups, viz., bacilli, cocci, spirilla, vibrios, spirochetes, etc. The bacteria have a thick cell wall of peptidoglycans. These are differentiated into gram positive and gram negative according to their behavior to the crystal violet and iodine solution, i.e., gram stain (La Placa 2005).

Bacterial communities involved in biodeterioration of monuments, mural, easels, and paintings on canvas can be divided into three nutritional groups. These are photoautotroph (use light as source of energy and carbon dioxide as a source of carbon), chemolithoautotrophs (use energy released from the oxidation of inorganic substances and carbon dioxide as a source of carbon), and chemoorganotrophs (use energy released from the oxidation of organic substances). Many species of cyanobacteria and sulfur-oxidizing and nitrifying bacteria produce strong acids like sulfuric acid and nitric acid, respectively. Chemolithoautotrophic endolithic nitrifying bacteria are reported to be present on the building stones. The role of nitrifying bacteria in deterioration of artworks was first reported by Kauffmann (1952) and Wagner and Schwartz (1965). Major genera of bacteria involved in biodeterioration of paintings are affiliated to *Pseudomonas*, *Bacillus*, *Alcaligenes*, *Flavobacterium*, and *Staphylococcus* (Nuhoglu et al. 2006; Capodicasa et al. 2010; Lopez-Miras et al. 2013; Pavic et al. 2015). The colonization of bacteria leads to the colonization of artwork by common fungi such as *Aspergillus*, *Penicillium*, *Cladosporium*, and *Alternaria* species (Gorbushina et al. 2004; Caselli et al. 2018), which are linked to the mechanical damages and/or paint discoloration due to their metabolic activities (Soffritti et al. 2019).

Easel paintings are deteriorated by the bacterial species affiliated to genera *Pseudomonas, Bacillus, Alcaligenes*, and *Flavobacterium* (Ciferri 1999). The bacterial communities lead to the further colonization by the fungi like *Aureobasidium* (*Pullularia*) *pullulans* as the main biodeteriogen.

Predominant cultivable bacterial species with their potential of degrading organic and inorganic components of easel and canvas paintings were affiliated to the genera, viz., *Bacillus* and *Staphylococcus*. Genera like *Staphylococcus*, *Acinetobacter*, *Agrococcus*, *Janibacter*, *Rhodococcus*, and *Stenotrophomonas* are cultured and isolated on artificial media from the deteriorated canvas. Most of the isolated bacteria produce an array of enzymes like esterase, proteases, and lipases which are involved in the deterioration of different organic components in the canvas, preparatory, and paint layers. Some bacterial strains have also been reported to produce endocellulase, responsible for degradation of cellulose in the support layers. This indicates that the isolated bacteria from the paintings are well equipped to harm the artwork as a whole (Pavic et al. 2015).

Lopez-Miras et al. (2013a, b) studied the *Cristo de la Paciencia*, an oil painting on canvas, which showed biodeterioration on the reverse side. Only four bacterial strains belonging to two genera of phylum *Firmicutes*, i.e., *Bacillus* and *Sporosarcina*, were isolated from the deteriorated portion of the painting although no fungal strains could be cultivated. The oil painting on canvas *Virgen de Guadalupe* was studied by Lopez-Miras (2013) using advanced biotechnological techniques RAPD analysis and comparing the sequences with databases and observed *Brevibacterium* sp. from the reverse face of the painting. This also clarified that *Brevibacterium* sp. and *Arthrobacter* have synergistic effect on each other. From different studies this may be concluded that the spore-producing bacteria such as *Bacillus* and fungi are the pioneers. Canvas is greatly damaged by cellulolytic bacteria and fungi; major bacterial genera are supposed to degrade the support material, e.g., *Cellvibrio*, *Sporocytophaga*, *Myxococcoides*, *Cellufalcicula*, and also *Clostridium* sp., as anaerobic bacteria has been reported (Caneva et al. 2002).

1.4 Conclusion and Future Prospects

On the basis of the analysis of literature available, it is concluded that the presence of microorganisms on or around a painting doesn't mean that all these are involved in the deterioration of paintings. These microorganisms indirectly affect the artwork by making the suitable habitat for microbial communities with a potential of biodeterioration. Sometimes they function in a synergistic manner. Although the reverse side of the canvas paintings are more susceptible to biodeteriogens, the colonization occurs more frequently on the exposed face perhaps due to the gravitational settlements. Most studies indicate that maximum communities on obverse face of oil paintings are inactive, while on the reverse side the communities observed are generally involved in biodeterioration. The study and identification of microbial communities of the reverse side of canvas paintings are less. This is a thrust field to explore the microorganisms, their metabolism, and the metabolic products responsible for biodeterioration of canvas paintings of cultural heritage and to develop the strategies for conservation and restoration of this kind of artworks.

References

Albertano P, Moscone D, Palleschi G, Hermosin B, Saiz-Jimenez C, Sanchez-Moral S, Hernandez-Marine M, Urzi C, Groth I, Schroeckh V, Saarela M, Mattila-Sandholm T, Gallon JR, Graziottin F, Bisconti F, Giuliani R (2003) Cyanobacteria attack rocks (CATS): control and preventive strategies to avoid damage caused by cyanobacteria and associated microorganisms in Roman hypogean monuments. In: Saiz-Jimenez C (ed) Molecular biology and cultural heritage. Balkema, Lisse

Alexander M (1981) Biodegradation of chemicals of environmental concern. Science 211:132–138

Caneva G, Nugari MP, Salvadori O (1991) Environmental factors in biodeterioration. In: Caneva G, Nugari MP, Salvadori O (eds) Biology in the conservation of works of art. ICCROM—Sintesi Grafica s.r.l, Rome, pp 3–24

Caneva G, Nugari MP, Salvadori O (2002) La Biologia de la restauración. Ed: Nerea SA, Espana. ISBN: 8489569487

Capodicasa S, Fedi S, Porcelli AM, Zannoni D (2010) The microbial community dwelling on a biodeteriorated 16th century painting. Int Biodeterior Biodegrad 64:727–733

Caselli E, Pancaldi S, Baldisserotto C, Petrucci F, Impallaria A, Volpe L, D'Accolti M, Soffritti I, Coccagna M, Sassu G (2018) Characterization of biodegradation in a 17th century easel painting and potential for a biological approach. PLoS One 13:e0207630

Ciferri O (1999) Microbial degradation of paintings. Appl Environ Microbiol 65:879–885

Dhawan S, Agrawal OP (1986) Fungal flora of miniature paintings and lithographs. Int Biodeterior Bull 22:95–99

Di Carlo E, Barresi G, Palla F (2017) Biodeterioration. In: Palla F, Barresi G (eds) Biotechnology and conservation of cultural heritage. Springer, Cham, pp 1–30

Gallo F (1985) Biological factors in deterioration of paper. Rome: ICCROM

Garg KL, Jain KK, Mishra A (1995) Role of fungi in the deterioration of wall paintings. Sci Total Environ 167:255–271

Gonzalez JM, Saiz-Jimenez C (2005) Application of molecular nucleic acid-based techniques for the study of microbial communities in monuments and artworks. Int Microbiol 8:189–194

Gorbushina AA, Heyrman J, Dorniedena T, Gonzalez-Delvalle M, Krumbein WE, Laiz L (2004) Bacterial and fungal diversity and biodeterioration problems in mural painting environments of St. Martins church (Greene–Kreiensen, Germany). Int Biodeterior Biodegrad 53:13–24

Guglielminetti M, De Giuli Morghen C, Radaelli A, Bistoni F, Carruba G, Spera G (1994) Mycological and ultrastructural studies to evaluate biodeterioration of mural paintings. Detection of fungi and mites in Frescos of the monastery of St Damian in Assisi. Int Biodeterior Biodegrad 33:69–83

Hirsch P, Eckhardt FEW, Palmer RJ Jr (1995) Fungi active in weathering of rock and stone monuments. Can J Bot 73:S1384–S1390

Hoffmann L (2002) Caves and other low-light environments: aerophytic photoautotrophic microorganisms. In: Bitton G (ed) Encyclopedia of environmental microbiology. John Wiley, New York, pp 835–843

Karamanos Y (1997) Endo-IV-acetyl-P-D-glucosaminidases and their potential substrates: structure/function relationships. Res Microbiol 148:661–671

Kauffmann MJ (1952) Rôle des bactéries nitrificantes dans l'altération des pierres calcaires des monuments. Competes Rendus Acad Sci 234:2395–2397

La Placa M (2005) Principi di microbiologia medica. Esculapio Editore, Bologna, Italy

Laiz L, Pinar G, Lubitz W, Saiz-Jimenez C (2003) Monitoring the colonisation of monuments by bacteria: cultivation versus molecular methods. Environ Microbiol 5:72–74

Lamenti G, Tomaselli L, Tiano P (2003) Cyanobacteria and biodeterioration of monumental stones. In: Saiz-Jimenez C (ed) Molecular biology and cultural heritage. Balkema, Lisse, pp 73–78

Leonardi R (2005) Nuclear physics and painting: sub-topic of the wide and fascinating field of science and art. Nucl Phys A 752:659–674

Lopez-Miras M, Pinar G, Romero-Noguera J, Bolivar-Galiano FC, Ettenauer J, Sterflinger K (2013a) Microbial communities adhering to the obverse and reverse sides of an oil painting on canvas: identification and evaluation of their biodegradative potential. Aerobiologia 29:301–314

Lopez-Miras MDM, Martın-Sanchez I, Yebra-Rodrıguez A, Romero-Noguera J, Bolıvar-Galiano F, Ettenauer J (2013b) Contribution of the microbial communities detected on an oil painting on canvas to its biodeterioration. PLoS One 8:e80198

Makies F (1981) Enzymatic consolidation of paintings. ICOM Committee for Conservation. 6th Triennal Meeting. Ottawa, 21–25 September 1981. Preprints. ICOM. 81/2/7-1–7n, Paris

Matteini M, Mazzeo R (2009) Structure of panel and canvas paintings. In: Pinna D, Galeotti M, Mazzeo R (eds) Scientific examination for the investigation of paintings, a handbook for conservator-restorers. Firenze, Centro Di, pp 11–20

Mihajlovski A, Gabarre A, Seyer D, Bousta F, Di Martino P (2017) Bacterial diversity on rock surface of the ruined part of a French historic monument: the Chaalis abbey. Int Biodeterior Biodegrad 120:161–169

Nuhoglu Y, Oguz E, Uslu H, Ozbek A, Ipekoglu B, Ocak I, Hasenekoglu I (2006) The accelerating effects of the microorganisms on biodeterioration of stone monuments under air pollution and continental-cold climatic conditions in Erzurum, Turkey. Sci Total Environ 364:272–283

Onions AHS, Allsop D, Eggins HOW (1981) Smith's introduction to industrial mycology, 7th edn. Edward Arnold, London

Pavic A, Ilic-Tomic T, Pacevski A, Nedeljkovic T, Vasiljevic B, Moric I (2015) Diversity and biodeteriorative potential of bacterial isolates from deteriorated modern combined-technique canvas painting. Int Biodeterior Biodegrad 97:40–50

Pellerito C, Di Marco AE, Di Natale MC, Pignataro B, Scopelliti M, Sebastianelli M (2016) Scientific studies for the restoration of a wood painting of the Galleria Interdisciplinare Regionale della Sicilia-Palazzo Mirto di Palermo. Microchem J 124:682–692

Peterson K, Klocke J (2012) Understanding the deterioration of paintings by microorganisms and insects. In: Stoner JH, Rushfield R (eds) Conservation of easel paintings. Routledge, New York, pp 693–709

Pinar G, Ramos C, Rolleke S, Schabereiter-Gurtner C, Vybiral D, Lubitz W (2001) Detection of indigenous *Halobacillus* populations in damaged ancient wall paintings and building materials: molecular monitoring and cultivation. Appl Environ Microbiol 67:4891–4895

Pinna D, Salvadori O (2008) Processes of biodeterioration: general mechanisms. In: Caneva G, Nugari MP, Salvadori O (eds) Plant biology for cultural heritage: biodeterioration and conservation. The Getty Conservation Institute, Los Angeles, pp 15–34

Poyatos F, Morales F, Nicholsen AW, Giordano A (2017) Physiology of biodeterioration on canvas paintings: physiology and biodeterioration. J Cell Physiol 233:2741

Radaelli A, Paganini M, Basavecchia V, Elli V, Neri M, Zanotto C (2004) Identification, molecular biotyping and ultrastructural studies of bacterial communities isolated from two damaged frescoes of St Damian's Monastery in Assisi. Lett Appl Microbiol 38:447–453
Ranalli G, Zanardini E, Sorlini C (2009) Biodeterioration—including cultural heritage. In: Schaechter M (ed) Encyclopedia of microbiology. Elsevier, Oxford, pp 191–205
Ravikumar HR, Rao SS, Karigar CS (2012) Biodegradation of paints: a current status. Indian J Sci Technol 5:1977–1987
Roldan M, Clavero E, Castel S, Hernandez-Marine M (2004) Biofilms fluorescence and image analysis in hypogean monuments research. Algologic Stud 111(1):127–143. https://doi.org/10.1127/1864-1318/2004/0111-0127
Rosado T, Silva M, Dias L, Candeias A, Gil M, Mirao J (2017) Microorganisms and the integrated conservation-intervention process of the renaissance mural paintings from Casas Pintadasin Evora—know to act, act to preserve. J King Saud Univ Sci 29:478–486
Rose AH (1981) Microbial biodeterioration. In: Economic microbiology. Academic Press, Ltd., London
Rubio RF, Bolivar FC (1997) Preliminary study on the biodeterioration of canvas paintings from the seventeenth century by *Microchiroptera*. Int Biodeterior Biodegrad 40:161–169
Saiz-Jimenez C (1993) Deposition of airborne organic pollutants on historic buildings. Atmos Environ 27:77–85
Salvador C, Bordalo R, Silva M, Rosado T, Candeias A, Caldeira AT (2017) On the conservation of easel paintings: evaluation of microbial contamination and artists materials. Appl Phys A Mater Sci Process 123:80
Sampa S, Luppi Mosca AM (1989) A study of the fungi occurring on 15th century frescoes in Florence, Italy. Int Biodeterior 25:343–353
Santos A, Cerrada A, Garcia S, San Andres M, Abrusci C, Marquina D (2009) Application of molecular techniques to the elucidation of the microbial community structure of antique paintings. Microb Ecol 58:692e702
Sassu G, Cappelletti F, Ghelfi B (2017) Carlo Bononi. L'ultimo sognatore dell'Officina ferrarese, 1st edn. Fondazione Ferrara Arte, Ferrara
Schabereiter-Gurtner C, Pinar G, Lubitz W, Rolleke S (2001) An advanced strategy to identify bacterial communities on art objects. J Microbiol Methods 45:77–87
Soffritti I, D'Accolti M, Lanzoni L, Volta A, Bisi M, Mazzacane S, Caselli E (2019) The potential use of microorganisms as restorative agents: an update. Sustainability 11:3853
Soliman NA, Knoll M, Abdel-Fattah YR, Schmid RD, Lange S (2007) Molecular cloning and characterization of thermo stable esterase and lipase from Geobacillus thermoleovorans YN isolated from desert soil in Egypt. Process Biochem 42:1090–1100
Souza-Egipsy V, Wierzchos J, Sancho C, Belmonte A, Ascaso C (2004) Role of biological soil crust cover in bioweathering and protection of sandstone in semi-arid landscape (Torrollones de Gabarda, Huesca, Spain). Earth Surf Process Land 29(13):1651–1661. https://doi.org/10.1002/esp.1118
Stulik D, Paint (2000) In: Taft WSJ, Mayer JW (eds) The science of paintings. Springer, New York, pp 12–25
Suihko ML, Alakom HL, Gorbushina A, Fortune I, Marquard J, Saarela M (2007) Characterization of aerobic bacterial and fungal microbiota on surfaces of historic Scottish monuments. Syst Appl Microbiol 30:494–508
Taft WSJ, Mayer JW (2000) The structure and analysis of paintings. In: Taft WSJ, Mayer JW (eds) The science of paintings. Springer, New York, pp 1–11
Tiano P (2002) Biodegradationof cultural heritage: decay mechanisms and control methods "9th ARIADNE Workshop" historic material and their diagnostic, ARCCHIP, Prague, 22–28 April 2002. http://www.arcchip.cz/w09/w09_tiano.pdf
Urzì C, Krumbein WE (1994) Microbiological impacts on the cultural heritage. In: Krumbein WE, Brimblecombe P, Cosgrove DE, Stainforth S (eds) Durability and change: The Science, Responsability and Cost of Sustaining Cultural Heritage. John Wiley & Son, Chichester, pp 107–135

Videla HA, Guiamet PS, De Saravia SG (2000) Biodeterioration of Mayan archaeological sites in the Yucatan Peninsula, Mexico. Int Biodeterior Biodegrad 46:335–341
Wagner E, Schwartz W (1965) Geomikrobiologische Untersuchungen. IV Untersuchungen über die mikrobielle Verwitterung von Kalkstein im Karst. Z Allg Mikrobiol 5(1):52–76. https://doi.org/10.1002/jobm.19650050108

Microbial Biocleaning Technologies for Cultural Heritage: Current Status and Future Challenges

2

Biswajit Batabyal

Abstract

Biocleaning technologies applied to cultural heritage sites have evolved to function in a wide range of environments, from laboratory conditions to cultural heritage monuments, stoneworks, frescoes and easel paintings. The accurate study of the microbial and fungal communities dwelling on artworks, and involved in their deterioration, is essential for the adoption of optimal prevention and conservation strategies. Biotechnologies have been able to resolve a range of problems on various artistic materials and to combat diverse artistic pathologies by using different cultures of viable bacteria. Microbial agents are among the major causes of deterioration of cultural heritage, strongly affecting our global cultural legacy, the stone, glass, wood and other sources used to fabricate millions of artworks and monuments around the world. Microbial action has been harnessed to clean the surfaces of stone sculptures and buildings and frescoes.

Keywords

Microbial agents · Biocleaning · Cultural heritage

2.1 Introduction

Microbial agents are among the major causes of deterioration of cultural heritage, strongly affecting our global cultural legacy, the stone, glass, wood and other sources used to fabricate millions of artworks and monuments around the world. Microbial action has been harnessed to clean the surfaces of stone sculptures and buildings and frescoes. In particular, the ability and potential of different microorganisms to remove undesired sulphates, nitrates and organic matter have

B. Batabyal (✉)
Serum Analysis Centre Pvt. Ltd., Kolkata, West Bengal, India

A. N. Yadav et al. (eds.), *Microbial Biotechnology Approaches to Monuments of Cultural Heritage*, https://doi.org/10.1007/978-981-15-3401-0_2

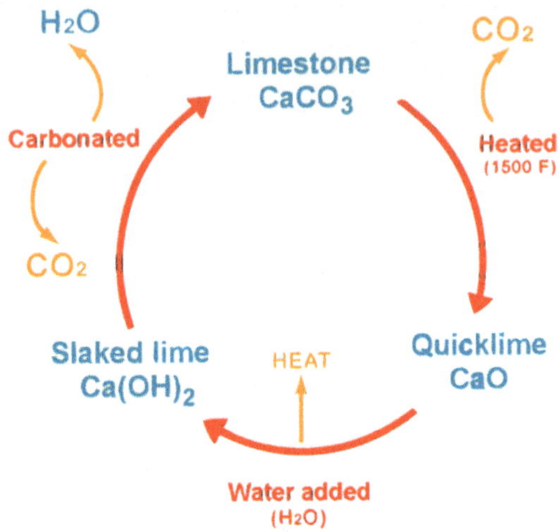

Fig. 2.1 Lime cycle

been demonstrated a number of times in the last decade. Microorganisms are generally associated with detrimental effects on artistic materials (Cappitelli et al. 2009, 2012). Microorganisms play important roles in biogeochemical cycles and in the food industry. Microorganisms can be used to clear out pollution (spilled oil, solvents, pesticides and other environmentally toxic compounds) in a process called bioremediation. The conservation and restoration of frescoes is the process of caring for and maintaining frescos and includes documentation, examination, research and treatment to ensure their long-term viability, when desired. The combined use of microscopy analysis and phenotypic/genotypic identification approaches can be usefully applied for the characterization of microbial diversity associated with biodeterioration processes in artworks, leading to the selection of the best strategy for restoration and prevention of recontamination.

2.2 Biocleaning Technologies

Fresco is a technique of mural painting in which pigment is applied to freshly laid or wet lime plaster (Fig. 2.1). Water acts as a type of binding agent that allows the pigment to merge with the plaster, and once the plaster sets the painting becomes an integral part of the wall.

Fresco Chemicals Consist of the Following:

- Silicon dioxide (sand)
- Calcium oxide (quick lime)
- Dihydrogen oxide (water)
- Calcium hydroxide (slaked lime)
- Carbon dioxide
- Calcium carbonate (limestone)

Calcium carbonate (limestone) is decomposed by heat to produce calcium oxide (quicklime) and carbon dioxide gas. Then calcium oxide reacts with water to form calcium hydroxide (slaked lime) which is accompanied by the release of heat, a reaction known as exothermic (Gardinali 2015).

2.2.1 Role of Sulphates

The biocleaning of altered surfaces could have some advantages over traditional cleaning methods as chemicals are not always selective and mechanical treatments can sometimes damage the surface. In addition, in the biological method, microorganisms act in the same way as they do in their natural environments. This preliminary experiment was successful but had two drawbacks: the treatment necessitated both the immersion of the object in a liquid medium and therefore was not suitable for large objects and the consolidation of the statue prior to the treatment. Additionally, the removal of gypsum was assessed only by visual observation. The limitation of method was the many days (10–14) required for the colonization of the sepiolite. Ion exchange chromatography analysis showed that bacteria entrapped in the Carbogel matrix allowed removal of 98% of sulphates from the marble, while Carbogel alone removed only a small percentage of the undesired salts. Biocleaning has been proven successful on many litho types, including limestone (Polo et al. 2010), even in the presence of heavy metals (Pedrazzani et al. 2006).

The biological procedure resulted in a more homogeneous removal of the surface deposits and preserved the patina noble under the black crust (Cappitelli et al. 2007). More recently, a study has evaluated the most appropriate cleaning treatment for black crust removal, adopting chemical (ammonium carbonate poultice), laser (1064 nm, Nd:YAG laser) and microbial cleaning for the removal of black crusts on colour artistic litho types of the Cathedral of Florence (Gioventù et al. 2011). The effects of the different procedures on the original surfaces were evaluated by scanning electron microscopy coupled with energy dispersive X-ray spectroscopy, Fourier transform infrared spectroscopy and colour measurements. One year later, further colour measurements were made. It was found that chemical cleaning led to non-homogeneous crust removal and that for the extremely powdery and incoherent red marlstone, the preferred treatment was laser cleaning.

2.2.2 Role of Nitrates

The bio removal of nitrate salts was performed from the tuff stone surfaces of the twelfth century Matera Cathedral, which had been altered by nitrates, using nitrate-reducing bacteria. In order to establish the impact of the biocleaning on the treated surfaces, colour variation and mineralogical and chemical analyses were carried out to evaluate the nitrate removal and to monitor the long-term effects on the surface. Nitrate removal was successful, and 6 years later, in the areas that had been treated using bacteria, the nitrate concentration remained stable; no noteworthy presence of microflora and colour change were measured in comparison with an untreated control area (Alfano et al. 2011). As physical and chemical treatments (such as ion exchange resins) did not yield the desired cleaning results, the paintings were cleaned using *Pseudomonas stutzeri* with a reduction of 92% in nitrates assessed by ion chromatography (Bosch-Roig et al. 2013).

2.2.3 Role of Organic Matter

The process of detaching frescoes from walls prior to restoration calls for notable quantities of organic compounds (such as glue and casein) that are distributed on both the painted surface and at the back of the fresco. The fresco was detached from the wall by conservators who, during the conservation treatment, had also mixed a biocide to the glue (Mukhopadhyay 2005). This biocide prevented the growth of microorganisms but unfortunately, over time, has favoured polymerization posing difficulty in removing such organic compounds by traditional methods. Bacteria were excellent bio restoration agents on the fresco (Ranalli et al. 2005).

The presence of calcium carbonate was attributed to the carbonate nature of the plaster on which the paintings were executed, gypsum to the plaster sulphitation processes or as a result of capillary infiltration of the walls and calcium oxalate to the degradation of a superficially applied compound. *Stenotrophomonas maltophilia* was used for the removal of organic deposits, *P. koreensis* for the removal of phosphate and organic deposits and *Cellulosimicrobium cellulans* for the solubilization of carbonate and gypsum deposits. A new support matrix, Laponite, was employed as it was found both appropriate for conservation purposes and compatible with the microorganisms. An important advantage of using the selected strains was that due to the difficulty of thermally insulating the loggia, the applications were carried out at temperatures ranging between 6 and 37 °C. Despite this, no decrease in the efficiency of the cleaning treatment was reported. Paints produced in the twentieth and twenty-first centuries, including spray paints used for graffiti, often contain synthetic polymers. Current methods for the removal of graffiti include chemical and physical (including laser) approaches. Bioremediation has a great potential as a novel approach to graffiti removal (Sanmartín et al. 2014). Microorganisms have been proposed for the removal of synthetic polymers used as conservation materials and in paints from cultural heritage surfaces, e.g. nitrocellulose (Troiano et al. 2014).

Frescoes can be found in places of worship such as churches, ancient temples and tombs, as well as private residences and commercial establishments used for public entertainment. It is these environments and their pollutants that interact with the chemicals, organic and inorganic, utilized to create the frescoes and the pigments used that contribute to their aesthetic and structural deterioration. Additionally, wall paintings such as frescoes depending on the technique used possess a layered structure consisting of support, ground or paint layer. These constituents of wall paintings undergo deterioration physically, chemically or biologically. Although factors like moisture, salts and atmospheric pollution have generally been the main

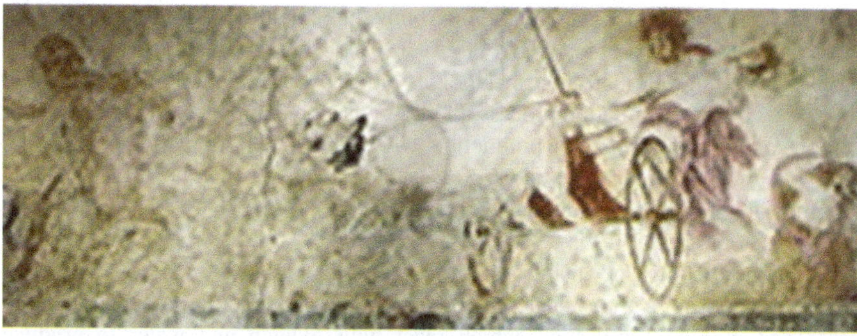

Fig. 2.2 Hades abducting Persephone

Fig. 2.3 Fresco in the terrace houses in Ephesus

contributors to the deterioration of wall paintings in most cases, many in the field believe the growth of biological agencies like fungi and microbial flora is also responsible for decay (Garg et al. 1995) (Figs. 2.2 and 2.3).

2.3 Chemical Degradation

The presence of pigment discolouration and stains and the formation of biofilm are indicative of chemical degradation. Given the variety of organic and inorganic molecules present in frescoes, many types of microorganisms may grow on the substrate of the fresco provided that environmental conditions (humidity, temperature, light and pH) are fostered (Ciferri 1999). Chemical deterioration can be attributed to fungi through their metabolites either by assimilation or dissimulation processes. In the assimilation process, the fungi use the components of frescoes as a carbon source through enzyme production, whereas in the dissimulation process, the decay is mainly by the excretion of waste products or secretion of metabolic intermediates including acids and pigments which can damage, stain or disfigure the surface (Garg et al. 1995).

Signs of cracking and disintegration of paint layers and the formation of paint blisters are indicative of physical/structural degradation. Industrial pollutants contain gases and burning fossil fuels which react with oxygen and water to produce sulphuric acid and nitric acids. These acids convert calcium carbonate (limestone) to calcium sulphate which becomes soluble in the water, and it forms large crystals within the surface layer causing the fresco to blister and flake off (Gardinali 2015). Aside from the adverse effects of environmental pollutants, fungal growth either on or below the surface can cause the dislodging of the paint layers further contributing to the physical and structural degradation of frescoes (Garg et al. 1995).

2.4 Preventive Care

Frescoes that have been removed from their original context and relocated to cultural institutions have the benefit of being in a more stable environment that is consistently monitored, even though they are at low risk. However, frescoes still at their place of origin, such as cultural heritage sites, are at high risk because they are vulnerable to environmental elements due to a high volume of tourist traffic in conjunction with other pollutants. Therefore, as with any similar object, data loggers are useful to monitor ambient conditions such as temperature and relative humidity, as well as thermohygrometric sensors for microclimate monitoring for fresco paintings in indoor, outdoor or semi-confined environments (Merello et al. 2012).

2.5 Cleaning Methods

Cleaning aims to restore artworks to how the artist intended them to look; however, how an artwork is cleaned will depend on the nature of the material to be removed. With paintings, a variety of organic solvents are used, but the most common solvent is water, often with chelating agents, surfactants or salts to control pH. Applying solutions through tissues, gels and sponges is becoming the norm, due to the level of control offered by holding the cleaning system at the upper surface of the art. Such gels, introduced in the late 1980s, are usually water-based emulsions thickened with cellulose or synthetic polymers. By slowly releasing the solvent, they prevent some of the swelling damage that free solvents cause to paint layers. During the 1960s, it became popular to use synthetic polymers to consolidate and stabilize frescoes – plaster-based wall paintings. They seemed like the perfect replacement for the wax coatings previously used, but over time it became clear that this was not the case. Their presence drastically changed the paintings' surface properties, causing mechanical stresses and crystallization of salts beneath the painting leading to accelerated disintegration. In addition, the polymers themselves became discoloured and brittle. By the mid-1990s, laser cleaning was established for stone and started to be used for other materials such as gilded bronzes and frescoes. A major breakthrough came when an Italian physicist at the National Research Council Institute of Applied Physics in Florence, Salvatore Siano, developed a method that used even shorter pulses, of only micro- to nanosecond duration (Brazil 2015). Another major innovation in the last decade is the use of colloid science and nanotechnology in conservation. In the mid-1990s, colloid scientist Piero Baglioni came up with a microemulsion: a clear mixture of organic solvent and water, stabilized with a surfactant that sits at the interface between the water and organic phases. Another unusual method of cleaning frescoes is with the use of specific types of bacteria to remove inorganic crusts and animal glues from frescoes. Because bacteria can produce a whole host of enzymes, they can deal with complex cleaning problems, metabolizing organic and inorganic matter into hydrogen sulphide, molecular nitrogen or carbon dioxide (Brazil 2015).

2.6 Repair and Restoration Techniques

During the eighteenth century, new techniques were perfected for the restoration and conservation of ancient works of art, including methods of detaching fresco paintings from walls. Detachment involves separating the layer of paint from its natural backing, generally stone or brick, and can be categorized according to the removal technique used. The oldest method, known as the a massello technique, involves cutting the wall and removing a considerable part of it together with both layers of plaster and the fresco painting itself. The stacco technique, on the other hand, involves removing only the preparatory layer of plaster, called the arriccio, together with the painted surface.

Finally, the strappo technique, without doubt the least invasive, involves removing only the topmost layer of plaster, known as the intonachino, which has absorbed the pigments, without touching the underlying arriccio layer. In this method, a protective covering made from strips of cotton and animal glue is applied to the painted surface. A second, much heavier cloth, larger than the painted area, is then laid on top, and a deep incision is made in the wall around the edges of the fresco. A rubber mallet is used to repeatedly strike the fresco so that it detaches from the wall. Using a removal tool, a sort of awl, the painting and the intonachino attached to the cloth and glue covering are then detached, from the bottom up.

The back of the fresco is thinned to remove excess lime and reconstructed with a permanent backing made from two thin cotton cloths, called velatini, and a heavier cloth with a layer of glue. Two layers of mortar are then applied: first a rough one and then a smoother, more compact layer. The mortars make up the first real layer of the new backing. The velatini cloths and the heavier cloth serve only to facilitate future detachments and are therefore known as the strato di sacrificio, or sacrificial layer. Once the mortar is dry, a layer of adhesive is applied and the fresco is attached to a rigid support made from synthetic material which can be used to reconstruct the architecture that originally housed the fresco. After the backing has completely dried, the cloth covering used to protect the front of the fresco during detachment is removed using a hot water spray and decoloured ethyl alcohol (Nocentini 2015).

Piero Baglioni has also pioneered the use of nanoparticles for repairing deteriorating frescoes. Artists generally painted directly onto wet calcium hydroxide plaster, which reacts with atmospheric carbon dioxide to form calcium carbonate (calcite). Over centuries, pollution and humidity cause the carbonate layer to break down, and sulphate, nitrate and chloride salts within the walls recrystallize, leading to deterioration of the painted surface. Baglioni was sure that nanoparticles would improve on conventional conservation methods. His treatment injects calcium hydroxide nanoparticles dispersed in alcohol, and their small size, just 10–100 nm, allows them to penetrate several centimetres into the frescoes and slowly reform the depleted calcite (Brazil 2015).

Antibiotics such as amoxicillin can be used to treat strains of bacteria living in a fresco's natural pigment which can turn them into powder (Avvisati and McGivern 2015). Another method of fresco repair is the application of a protection and support bandage of cotton gauze and polyvinyl alcohol. Difficult sections are removed with soft brushes and localized vacuuming. The other areas that are easier to remove (because they had been damaged by less water) are removed with a paper pulp compress saturated with bicarbonate of ammonia solutions and removed with deionized water. These sections are strengthened and reattached and then cleansed with base exchange resin compresses, and the wall and pictorial layer are strengthened with barium hydrate. The cracks and detachments are stopped with lime putty and injected with an epoxy resin loaded with micronized silica (Cacci et al. 2003).

2.6.1 Sistine Chapel

The Sistine Chapel was restored in the late 1970s and through the 1980s. This was one of the most significant, largest and longest art restoration projects in history. The entire project took 12 years to complete, not taking into account the inspections, planning and approval of the project. Among the many parts of the chapel that was restored, what drew the most attention, were Michelangelo's frescoes. The restoration sparked controversy. A number of experts criticized the proposed techniques, claiming that the restoration procedure would scrape off the layers of various materials on the frescoes, which would lead to damage beyond repair, and that the removal of the materials would expose the pigments on the frescoes which were fragile and dated to artificial light, temperature variations, humidity and pollution. Such exposure, they feared, would cause massive damage to the original artwork.

For the frescos of Villa of the Mysteries in Pompeii, early conservation efforts sometimes involved removing frescoes, rebuilding or reinforcing the walls and then reattaching the paintings. The first conservators also applied a coat of wax mixed with oil to clean the paintings' surfaces, preserve the ancient pigments and stabilize the fragile works, giving the frescoes a glossy appearance the ancient artists never intended them to have. At the same time, the wax filled in cracks in the surfaces, sealing moisture inside the walls, further weakening them by compromising the strength of the mortar holding the walls together. By 2013 the villa, like most of Pompeii, was in dire need of modern conservation, as was a protective covering that had been constructed in different phases throughout the years. Parts of paintings were crumbling from unstable walls, and the mosaics had been severely damaged by millions of visitors' feet. Repeated applications of wax had caused the pigments to oxidize and darken, and the frescoes to yellow, significantly altering their appearance. All the surface decorations of the villa, both mosaics and frescoes, had been conserved before, but in irregular ways. Some of the methods currently being employed have been used by decades of conservators at Pompeii. Frescoes have been cleaned by hand using a scalpel or a chemical solution. Painted surfaces have been consolidated with an acrylic resin diluted with deionized water and then injected into cracks (Lobell 2015) as well as the use of antibiotics for the removal of bacteria (Avvisati and McGivern 2015). The teams today also have more high-tech tools at their disposal, including lasers to clean the frescoes and ultrasound, thermal imaging and radar to evaluate the level of decay of the walls and paintings. Drones are being used to examine the entirety of the villa's protective covering.

2.7 Conclusion and Future Prospects

Biodiversity is not evenly distributed; rather it varies greatly across the globe as well as within regions. Among other factors, the diversity of all living things (biota) depends on temperature, precipitation, altitude, soils, geography and the presence of other species.

The effect of organic farming has been a subject of interest for researchers. Theory suggests that organic farming practices, which exclude the use of most synthetic pesticides and fertilizers, may be beneficial for biodiversity. This is generally shown to be true for soils scaled to the area of cultivated land, where species abundance is, on average, 30% richer than that of conventional farms. However, for crop yield-scaled land, the effect of organic farming on biodiversity is highly debated due to the significantly lower yields compared to conventional farms.

In ancient farming practices, farmers did not possess the technology or manpower to have a significant impact on the destruction of biodiversity even as mass production agriculture was rising. Nowadays, common farming methods generally rely on pesticides to maintain high yields. With such, most agricultural landscapes favour monoculture crops with very little flora or fauna coexistence. Modern organic farm practices such as the removal of pesticides and the inclusion of animal manure, crop rotation and multicultural crops provide the chance for biodiversity to thrive.

Microorganisms display wide diversity in enzyme production, including lipases, proteases and oxidoreductases. It is now recognized as being a viable alternative to traditional chemical treatments such as organic solvents or other aggressive conservation methods like mechanical treatments. Microorganisms are the new bioagents for the recovery and conservation of artwork and historical architectural monuments. The basic idea of these innovative biological methods (biocleaning, bio-consolidation) is encouraged by the fact that only a few known microorganisms play a destructive role (causing deterioration) in the natural processes, while the majority of them are responsible for 'virtuous' processes. In addition, microorganisms can have advantages over chemical methods and enzymes in cleaning cultural heritage (CH), especially when the substances to be cleaned are complex and encrusted, due to their specificity of a pool of enzyme production. The great biodiversity in existing microorganisms and the selection of natural microorganisms using microbiological techniques such as culture-dependent enrichment and/or culture-independent molecular methods permit the isolation of the appropriate microorganisms for biological processes for CH purposes without resorting to the use genetically modified organism (GMO), which could potentially, through diffusion of the biological techniques, lead to additional, unforeseen risks to safety.

Ready to use biocleaning product to the conservation market, two basic aspects must be taken into account: the microorganism itself and the delivery system to be used (Bosch-Roig et al. 2015). At present, the delivery system is generally chosen by conservators according to their experience. Nevertheless, the delivery system is one of the most important aspects of biocleaning technology to take into account some carrier features before using them: (a) the ability to provide the microorganisms with the right conditions (e.g. water), (b) easy applicability to all types of surfaces whatever the orientation and (c) easy preparation and application and also convenient elimination at the end of the treatment. Biocleaning proposed as the traditional chemical- or physics-based cleaning methods was not entirely satisfactory, with the latter displaying many drawbacks including health problems caused to conservators and the risk of further damaging the object. In contrast, biocleaning offers many advantages including selectivity, no effects on the health of conservators, no pollution, relatively low cost, high homogeneity of the deposits removal and absence of ethical implications (Vergès-Belmin 1996; Webster and May 2006). In conclusion, being a sustainable approach, biocleaning is a way to go green in conservation practice.

The importance must be paid to the safety of these biocleaning technologies for the cultural heritage, the restorers and the environment. Surfaces which have previously been cleaned and which are being cleaned and to be cleaned using biocleaning technologies must be monitored in order to confirm and validate the biocleaning process. Importance must be given to developing suitable strategies for inspection and to monitoring any new microbial interactions on biocleaned artworks. When possible, these should include adequate on-site technologies based on non-invasive tools to understand the potential risks on biocleaned tangible heritage and include physical-chemical, biological and aesthetic analyses. Similarly, further research and clear demonstrations of the complete safety of biocleaning are of fundamental importance because this technology has a very significant role to play in the introduction and diffusion of a new approach to the application of human-friendly, environmentally sustainable techniques and technologies for the conservation and restoration of heritage properties. Only through growing awareness of this philosophy will the preservation of the cultural heritage left by our ancestors not occur at the expense of degradation, as is sometimes sadly happening currently through the use of traditional toxic organic solvents and aggressive techniques and products.

The confirmation of the absence of risk of these bio-application methodologies will permit their diffusion and application for the removal of several different types of alterations on a variety of artistic materials around the world (paints on wood, textiles, paper, papyrus and so on, as in addition to stone materials and frescoes). Very easily applied, ready-to-use biocleaning products, that include fast application and removal from altered surfaces, will finally become a reality, when the above-mentioned doubts about safety issues have successfully been clarified. At that time appropriate cost-to-benefit evaluation will confirm biocleaning technology to not only be environmentally sustainable but also economically viable. An expansion in this field and further development of biotechnological techniques will open up new opportunities to biologists for the conservation and restoration of cultural heritage sites.

References

Alfano G, Lustrato G, Belli C (2011) The bioremoval of nitrate and sulfate alterations on artistic stonework: the case-study of Matera cathedral after six years from the treatment. Int Biodeterior Biodegradation 65:1004–1011

Avvisati C, McGivern H (2015) Pompeian Frescoes cured with antibiotics. The Art Newspaper

Bosch-Roig P, Regidor-Ros JL, Montes-Estelles R (2013) Biocleaning of nitrate alterations on wall paintings by Pseudomonas stutzeri. Int Biodeterior Biodegradation 84:266–274

Bosch-Roig P, Lustrato G, Zanardini E, Ranalli G (2015) Biocleaning of cultural heritage stone surfaces and frescoes: which delivery system can be the most appropriate? Ann Microbiol 65:1227–1241

Brazil R (2015) Conservative innovations, Chemistry world. Royal Society of Chemistry

Cacci, ed Leonardo, La Fenice (eds) (2003) A building site. Marsilio, Venezia, p 118

Cappitelli F, Toniolo L, Sansonetti A, Gulotta D, Ranalli G, Zanardini E, Sorlini C (2007) Advantages of using microbial technology over traditional chemical technology in removal of black crusts from stone surfaces of historical monuments. Appl Environ Microbiol 73:5671–5675

Cappitelli F, Abbruscato P, Foladori P, Zanardini E, Ranalli G, Principi P, Villa F, Polo A, Sorlini C (2009) Detection and elimination of cyanobacteria from frescoes: the case of the St. Brizio Chapel (Orvieto Cathedral, Italy). Microb Ecol 57:633–639

Cappitelli F, Salvadori O, Albanese D, Villa F, Sorlini C (2012) Cyanobacteria cause black staining of the National Museum of the American Indian building, Washington, DC, USA. Biofouling 28:257–266

Ciferri O (1999) Microbial degradation of paintings. Appl Environ Microbiol 65:879–885

Gardinali PR (2015) Chemistry and Fresco painting. F I U

Garg KL, Kamal J, Mishra AK (1995) Role of fungi in the deterioration of wall paintings. Sci Total Environ 167:255–271

Gioventù E, Lorenzi PF, Villa F (2011) Comparing the bioremoval of black crusts on colored artistic lithotypes of the Cathedral of Florence with chemical and laser treatment. Int Biodeterior Biodegradation 65:832–839
Lobell JA (2015) Saving the villa of the mysteries. Archaeology Magazine. Archaeology Institute of America
Merello P, Garcia-Diego F, Zarzo M (2012) Microclimate monitoring of Aridane's house (Pompeii, Italy) for preventive conservation of Fresco painting. Chem Cent J 6:145
Mukhopadhyay R (2005) Bacteria rescue fresco. Anal Chem 77:457A–458A
Nocentini S (2015) 'Strappo' Detachment. Museo Benozzo Gozzoli
Pedrazzani R, Alessandri I, Bontempi E (2006) Study of sulphatation of Candoglia marble by means of micro X-ray diffraction experiments. Appl Phys A Mater 83:689–694
Polo A, Cappitelli F, Brusetti L, Principi P, Villa F, Giacomucci L, Ranalli G, Sorlini C (2010) Feasibility of removing surface deposits on stone using biological and chemical remediation methods. Microb Ecol 60:1–14
Ranalli G, Alfano G, Belli C, Lustrato G, Colombini MP, Bonaduce I, Zanardini E, Abbruscato P, Cappitelli F, Sorlini C (2005) Biotechnology applied to cultural heritage: biorestoration of frescoes using viable bacterial cells and enzymes. J Appl Microbiol 98:73–83
Sanmartín P, Cappitelli F, Mitchell R (2014) Current methods of graffiti removal: a review. Constr Build Mater 71:363–374
Troiano F, Vicini S, Gioventù E (2014) A methodology to select bacteria able to remove synthetic polymers. Polym Degrad Stab 107:321–327
Vergès-Belmin V (1996) Towards a definition of common evaluation criteria for the cleaning of porous building materials: a review. Sci Technol Cult Herit 5:69–83
Webster A, May E (2006) Bioremediation of weathered-building stone surfaces. Trends Biotechnol 24:255–260

Role of Bacterial Communities to Prevent the Microbial Growth on Cultural Heritage

3

Hina Upadhyay, Vandna Chhabra, and Jatinder Singh

Abstract

Different microorganisms have a considerable and important role in maintenance and prevention of cultural heritage buildings or different materials from deterioration. They have key role in deterioration processes. So, there is an urgent need to upgrade and apply new technologies and methods to identify the responsible microorganisms and manage them by using eco-friendly methods. However, there is a common consent that microorganisms play a significant role in the deterioration of cultural heritage, along with the same few microbes also play a significant role as a biocleaning agents, for the conservation of such type of materials. Recent research is carried out in this field where few microbial especially bacterial strains have been utilized as bio cleaning agents, which are frequently used in the preservation and restoration of cultural heritage. In this book chapter, we are highlighting the concerned common deteriorating agents and its control by using different strategies, with reference to the role of various kinds of bacterial communities for the management of such kind of problems in historical-cultural materials.

Keywords

Biodeterioration · Cultural heritage · Sulphate-utilizing microbes · Biocleaning technique

H. Upadhyay · V. Chhabra
Department of Agronomy, School of Agriculture, Lovely Professional University, Jalandhar, India

J. Singh (✉)
Department of Horticulture, School of Agriculture, Lovely Professional University, Jalandhar, India
e-mail: jatinder.19305@lpu.co.in

A. N. Yadav et al. (eds.), *Microbial Biotechnology Approaches to Monuments of Cultural Heritage*, https://doi.org/10.1007/978-981-15-3401-0_3

3.1 Introduction

In the world, different monuments, sculptures, wall paintings (paintings) and wood carving are scattered over various zones, showing the richness of our cultural heritage. Different kinds of insects, microorganism and weather agents are major responsible factors for their deterioration. Broadly responsible factors can be categorized into three different agents, i.e. physical agents, chemical agents and biological agents.

Nevertheless, the prime and significant damage is done by biological agents (microorganisms, etc.) (Tilak and Pande 1997). Under these conditions, the understanding of cultural property and phenomenon of biodeterioration becomes important for us. Cultural heritage or property is a term coined by UNESCO which includes archaeology, museum materials, architecture, artistic work, historical buildings and mural paintings and ethnographical objects stored in such type of places, while biodeterioration can be defined as any kind of variation, change or alteration in these objects brought about by above-mentioned agencies. This process of biodeterioration may result in spoilage of objects or staining or damaging of the same.

An entirely new field of research for the application of aerobiology has emerged in relation to biodeterioration of materials in archives and museum objects like statues, paper materials, wooden objects, leather material and other irreplaceable objects of historical importance. In every type of cultural property, different kinds of materials are used which decide for shape of the object. In biodeterioration process, different kinds of microorganisms are included like fungi, algae, bacteria, insects, birds, etc. One similarity occurs in these agencies that they are common under hot and humid conditions than cold and dry conditions. There are a lot of research opportunities in this field as the loss caused by different organisms are responsible as a biodeterioration agents to our cultural properties. Very less or scare scientific studies are available in literature related to current topic. Henceforth more and deep studies are required in this field. To control the damage caused by biodeteriorating agents, there are many challenges, as how to safely and economically solve such type of situation.

Inside museum and libraries, growth and development and multiplication of fungus is the main and serious problem. Under such conditions, brown spots called foxing are very frequent visual. Firstly, it was thought that these spots were common only on cellulosic materials. But we can also observe them on stony material along with insect infestation challenges. It is also studied by Hideo Arai (Hideo Arai 2013) that due to excess humidity and absence of ventilation causes of blackening of stony material.

A considerable loss has resulted because of biodeterioration, and above all, the loss of cultural and artistic materials is irreversible in nature. Damage and spoilage of cultural property is a common phenomenon in various parts of the world. Under such circumstances, various agencies of biodeterioration like environment factors including light, temperature, humidity, dust or dirt play significant role. Besides these, termites also cause heavily damage especially to wood works of these buildings.

The major problem faced in this respect is the analysis of responsible agents and deterioration process in nature. Under conventional systems, visual identification including microscopic examination and the isolation and identification of causal agents are some of the common steps to be followed. But the use of laboratory not only adds cost to the method but is also time-consuming. To carry such studies concerning biodeterioration components, conservation aspect should be given utmost importance. More and intensive studies are to be carried out to conserve and control damage of cultural property.

3.2 Common Organisms Involved in the Biodeterioration

Various kinds of microorganisms can proliferate on CH due to various reasons like closed environment of castles, caves, museums, libraries and churches, but it is a very important fact that the source of microorganisms may be from outside atmosphere or also within particular site or a result of human activities. For instance, these microbes may instigate via the skin and respiration-related activities which can further their development and multiplication (Sand 2001). Proper understanding of procedure along with different approaches for long-lasting control is mandatory for prevention of such situation and resultant biodeterioration (Castanier et al. 1999).

It is very true under the circumstances when relative humidity and temperature become suitable for their growth and development, such as during prevalence of hot weather along with humid conditions >65–70% (De Muynck et al. 2010; De Belie and Wang 2016).

Some microorganisms were also used to eradicate noxious and harmful chemical compounds from HC objects like sulphate, etc. No doubt, the use of suitable microorganism to preserve heritage is a good technique, but it has given rise to many logical questions. Today restorers from public and private sector are seeking more and more information from research scholars about the safety of this technology. But before proceeding, previously cleaned HC, being cleaned HC and to be cleaned HC by these technologies must be scrutinized in order to validate the process of biocleaning. To verify safe use of this technology and long-lasting effect, already work done by this technique should be subjected to various environmental conditions.

3.3 Type of Damage Identification

3.3.1 Cultural Heritage and Damage Type

In India, there are a number of historical buildings, among them few famous heritage buildings, the Taj Mahal , the Red Fort, Qutub Minar, etc. Furthermore, Ajanta and Ellora Caves and Charminar are also important monuments of cultural heritage. It is clear and well understood that these all are buildings, so they have to face the outside harsh weather as well influence of biodeteriorating agents and also have to go through the process of worsening. The damage to all this type of monuments may start from a single crack in its structure and loosening of building blocks mainly from walls of roof and through leakage and seepage of water and fading of colours along with corrosion. Warscheid et al. (2000) also said that in all these HCs, different kinds of stone material are used, and the same ranged from small statues to huge building. All these monuments are huge stony and concrete materials and have to face every type of weather action (physical, chemical and biological factors) (Verma et al. 2015). These events are historical events that are preserved through different kinds of monuments.

They include examples like the Great Pyramids (Egypt), Eiffel Tower (Paris), Westminster Abbey (London) and Statue of Liberty (USA) (González-Ruibal 2005). Due to various nature conservation challenges, collaborative efforts are expected between scientists and conservators. Degradation process or weathering of rocks is a well-known and common phenomenon. It is essential for life evolution, and it has been occurring since earth has emerged, and till date it is going and will continue to the future. Conversely, this loss is irreversible in nature and also includes decaying of traditionally meaningful stone objects (Kumar and Kumar 1999). Different kinds of physical factors like water, wind, temperature and moisture as well as chemical factors have strong influence on the outer surfaces of these monuments especially stones.

They can cause chemical reactions, oxidation and hydration, and can also affect the stability of rock matrix, solubilization process ultimately corrosion of materials that forms the stone and dissolution of carbonates (Koestler 2000). Furthermore, other anthropogenic factors (e.g. different sources of air pollution like smoke from vehicles, fumes of gases and others) may lead to worsening of this condition by increasing the concentration of chemical salts and compounds and their rate of deposition on stony materials. These factors contribute in a very significant and important way (Horie 1987).

3.3.2 Type of Conservation Treatment

Some microorganisms were also used to eradicate noxious and harmful chemical compounds from HC objects like sulphate, etc. To restore and preserve our cultural heritage, biotechnology has huge and important untapped potential. To protect and consolidate *these* CH and monuments, various types of treatments and efforts are tried before the occurrence of extensive loss (granular disintegration) which may lead to serious loss of surface material, and above all, this damage is irreversible in nature (Lewin and Baer 1974). The word consolidation means hardening of friable material and decaying the material. It can be done with the help of hardening product or by using cement, while protection means to protect the object by making it waterproof or to stop the entry of moisture in its core.

In nutshell we can say that the entry of different weathering agents is prevented. These kinds of treatments were common and frequently used during earlier times with the help of organic and inorganic materials (Price et al. 1988), for example, Ba $(OH)_2$ solutions and acrylic or epoxy resins (Price et al. 1988). Nevertheless, till date no satisfactorily treatment is available which can lead to 100% results. When different kinds of organic treatments are applied on these materials sometime there is a formation of incomplete harmful surface film on them. Moreover, they may also emit harmful, toxic solvents or gases which can cause health problems. Inorganic type of alliance is also preferred as stony materials (minerals) share some physico-chemical affinity (Tiano et al. 1999). For example, in case of lime water treatment, there is existence of calcium hydroxide solution that probably helps to combine carbonate stones as $Ca(OH)_2$ straightforwardly carbonates in presence of Co_2, which is present in atmosphere. At the end of reaction, there is a formation of CaCO3 compound. Nevertheless, the use of lime water treatment results in formation of very thin (micrometre-thick), superficial, friable submicron-sized calcite particles and results in partial protection (McNamara et al. 2006). Due to textural characteristics and compositional nature, different kinds of limestones, marbles and dolostones are very vulnerable to weathering process. As this weathering cycle continues, the porosity occurs in these stony surfaces and eventually decreases its mechanical strength of such type of material.

Numerous kinds of conservation treatments are applied to strengthen or modify stone features so that process of decaying can be minimized. A diverse and large number of microorganisms that can colonize at surface of the stone as well as inside of the stone are supported by these historic stones (Ghosh et al. 2006). Several species of microorganism partake in process of mineral carbonate precipitation in various environments (geological formations, saline lakes and oceans).

Nowadays use of specific kinds of microorganism which may yield microbial mineral precipitation resulting from metabolic activities and use of biomineralogy concept may result in potential invention of a new salt. The property of bacteria concerning self-repairing is remarkable, and it may be used further to manage the cracks and crevices in concrete. Worldwide our cultural bequest is being spoiled by various types of microbial agents. Different types of materials like the stones, wood and fabrics are used to fabricate artworks including monuments in the world. Action by different microbes has been harnessed for the purpose of cleaning of stony materials like sculptures and buildings. Since the last decade, the ability and probability of several microorganisms to manage different chemical compounds like nitrates and sulphates including organic matter have been described many times. In addition, these microorganisms can be used in the bioremediation process to clear pollution particularly in the case of pesticides, spilled oil and other environmentally toxic compounds.

It has been reported earlier that various microorganisms have harmful effects on artistic materials (Law and Slepecky 1961). Microorganisms have significant role in food industry and in biogeochemical cycles. For instance, various kinds of fermentative activities of yeast are mandatory to give rise to CO2 which result in dough, alcohol etc.

3.4 Current Scenario Concerning Use of Microorganisms in Preservation of Cultural Heritage

Since ancient times, different types of microorganisms are being used in various sectors like healthcare, energy production, food production industry, treatment of wastewater and agriculture. But the use of microorganism has started, i.e. biocleaning, in the 1990s, and since then, their use has been increasing day by day. Today it has emerged as better alternative compared to other traditional chemical treatments like mechanical treatments or organic solvents, etc. Microorganisms are bioagents which are frequently used in conservation of artwork and historical architectural monuments.

The prime idea behind this innovation (biocleaning, bioconsolidation) is that all microorganisms are not harmful (causing deterioration); only a few of them cause such kind of activities. Most of microorganisms may lead to virtuous processes. There is a lot of variation as far as biodiversity is concerned among microorganisms including bacteria, eukaryotes and archaea. Furthermore, they are surprisingly diverse concerning their development requirements, and their rate of multiplication is particularly influenced by the availability of nutrients in their surrounding environments, although their current requirements are almost the same, that is. Sources of energy and various types of nutrients, including macronutrients to detect elements such as Zn, Mn, Co and Cu, and water. Hence it may be concluded that they have some common factors for survival like optimum temperature, water and air, pH and salinity.

Under such conditions, the role of biotechnology can't be ignored as by using different cultures of bacteria, various problems related to wall paintings, marble statues, etc. can be managed easily. Use of suitable bacteria culture greatly combats against these artistic pathologies like black crust and removal of other organic substances. In literature various types of bacteria and their role to is already mentioned, Sulphur ion, *Desulfovibrio vulgaris* is an example of sulphate-reducing bacteria and nitrate-reducing bacteria like *Pseudomonas stutzeri* (Gauri et al. 1992; Ranalli et al. 2000, 2005). However, in nature, a lot of microbial communities are present which have positive influence and can serve humanity in a better way (Table 3.1).

3.5 Certain Chemical Compounds, Role of Microbes and Literature Available About Managing Them

Gauri et al. (1992) were pioneers in using microorganism particularly sulphate-reducing bacteria for the purpose of removal of black layer from old gypsum, marble statue. They have used sulphate-reducing bacteria (*Desulfovibrio desulfuricans*) for this purpose. This treatment was based on the concept that biocleaning methods have some additional benefits over traditional cleaning methods (especially chemicals), and they are not always selective; moreover, sometimes use of mechanical methods may also prove to be harmful. It is supposed that suitable microorganism may act in

Table 3.1 Effective bacteria and their role in the conservation treatment of monuments and artwork

Sr. no.	Organism	Type of work	References
1	*Desulfovibrio desulfuricans*	Removal of black crust from marble surface	Gauri et al. (1992)
2.	*Pseudomonas fluorescens*	Biocleaning and restoration of fresco, calcite precipitation	Anderson et al. (1992)
3.	*Pseudomonas* and *Bacillus* sp.	Removal of phenanthrene deposit from weathered stones	Sáiz-Jiménez (1997)
4.	*Bacillus cereus*	Biocement (Biocalcin) formation and limestone consolidation	Orial et al. (1992)
5.	*Pseudomonas* sp.	Removal of nitrates from the weathered stones	Ranalli et al. (2000)
6.	*Myxococcus xanthus*	Consolidation of ornamental stones and bioconservation of cultural heritage structures	Rodriguez-Navarro et al. (2003) and Jroundi et al. (2010)
7.	*Halobacillus trueperi*	Biomineralization	Rivadeneyra et al. (2004)
8.	*Pseudomonas cepacia*	Biocleaning, restoration and removal of animal glue from fresco	Ranalli et al. (2005)
9.	*Bacillus megaterium*	Calcite precipitation	Cacchio et al. (2003)
10.	*Acinetobacter*	Limestone consolidation	Zamarreno et al. (2009)
11.	*Thiobacillus* sp.	Removal of fouled layer of lichen from weathered concrete specimens	De Graef (2005)
12	*Pseudomonas halophile*	Calcite precipitation	Rivadeneyra et al. (2006)
13.	*Desulfovibrio vulgaris*	Removal of black crust from marble surface	
14.	*Bacillus pasteurii*	Biocement formation and consolidation of sand column and repair of concrete cracks	Sarda et al. (2009)

similar way as they do under natural conditions. The very first experiment concerning use of microorganism was satisfactory, but dipping or immersion of object was mandatory in solution, but it was very difficult for huge objects. Consolidation was a prerequisite in such cases. After this treatment, as an improvement, objects were treated with *D. desulfuricans* and *D. vulgaris*. Two different objects were selected for this experiment, viz. an old marble “horse hoof” and an old marble column. Immersion was avoided, as inorganic material sepiolite was used as a supply source for bacteria. Moreover, only a few applications (three) are required for a short duration (for a total of 45 h). Through ion-exchange chromatography analysis, it is known that the bacteria in the carbogels recovered 98% sulphate compounds from stony material, but when carbogel is applied alone, results are very limited.

Biocleaning chemical treatment involved an ammonium carbonate-EDTA mixture, while in bio cleaning, there is involvement of *D. Vulgaris*, which is aimed for the elimination of black crust from stony material of the Milan Cathedral (Italy). From this experiment, it can be concluded that biological methods acted in more efficient way in removal of crust. Recently, a research study has been conducted to access most suitable cleaning process, aiming to remove black crust. Both chemical and microorganisms were used in this experiment. There was involvement of different chemicals like ammonium carbonate poultice and laser (1064 nm, Nd: YAG laser) along with microbial cleaning method on artistic lithotypes of the Cathedral of Florence.

Impacts of both treatments on artistic lithotypes were assessed by various methods, for example, scanning electron microscopy along with energy-dispersive X-ray spectroscopy and Fourier-transform infrared spectroscopy including colour measurements.

After a period of 1 year, colour measurements were again conducted. It was concluded that results were not uniform in case of chemical cleaning regarding crust removal, while for incoherent red marlstone, which was extremely powdery, laser treatment was used. The most satisfactory results were noted in case of microbial cleaning process. It was recorded as a manageable and competent process for removal of sulphate chemical. At the same time, it was observed the duration and frequency of treatment was too lengthy and high numbered to get successful and satisfactory results (almost four poultice time application and each for 10 h).

Addressed this time aspect and accessed the impacts of sulphate-reducing bacterium strain on artistic marble, influenced by black crusts. When comparison was made with biocleaning alone, it was concluded that it needs fewer biological applications and total cleaning time was reduced to 70%, still retentive all benefits of the biocleaning techniques.

3.5.1 Microbial $CaCO_3$ for Strength Improvement of Stony Monuments

Variable results were obtained by mixing bacteria in cement material, for strengthening purpose. It was reported by many researchers that rise of 28-day compressive strength with 9–25% was recorded when bacteria were mixed in concrete (Ghosh et al. 2005, Achal et al. 2012; Basaran 2013). Park et al. (2010) and Ghosh et al. (2005) also described both negative and positive impacts of bacteria mixing in concrete. Roughly 2% viability retention was noted of *S. pasteurii* cells in hardened cement paste, with age 330-day old (Basaran 2013), and it was further recorded that 50% cells were considered as metabolically active vegetative cells. However, a lot of problems are there to prove the active contribution of bacteria to calcium carbonate precipitation, inside concrete matrix at the high pH of 13. Another aspect is the feeding of selected bacteria to become active. For this purpose, mortar mix is added with required nutrients. Such addition of nutrients also influences the strength of mortar. Impacts of calcium nitrate (as source of calcium) and yeast extract (as nutrient) was evaluated by Wang et al. (2014), with isothermal calorimetry on the cement hydration. They concluded that Ca nitrate can speed up cement hydration, but at the same time, delayed hydration was observed in case of yeast extract (0.85% versus cement weight) at 7 days of age and lessened compressive strength at 3 months.

3.5.2 Microbial $CaCO_3$ for Surface Consolidation

First time, the tower of the Saint-Medard Church located in Thouars was applied with bioconsolidation treatment in 1993 (LeMetayer-Levrel et al. 1999). It was reported that biomineralization was also tried by Tiano et al. (2006) and Jroundi et al. (2010) for on-site preservation of decomposed stones. Strengthening impact of biodeposition was recorded by De Muynck et al. (2010) on Maastricht limestone through DRMS Cordless SINT Technology in Italy.

It was recorded that biodeposition strengthens limestone up to a distance of 30 mm (depth wise), but at the same time, the results were better or similar to that of traditional surface treatments. The research workers were able to cut the expanses of application, according to traditional consolidates. Nevertheless, in this regard, one major point is the application of hygroscopic salts (e.g. acetate, calcium chloride, etc.) aiming bioprecipitation, and this treatment can lead to damage to the stony material at later stages.

Till now, only HC buildings were bioconsolidate with the help of porous calcareous stones. However, many researchers have used MICP treatment on stony material with good success rate, but under laboratory conditions only including very few on-site tests. Stone is much denser compared to porous limestone. Being porous in nature, fewer bacteria are retained on the surface layer. Increased resistance was also recorded during thawing and freezing. Application of biodeposition leads to analogous results as conventional surface treatments.

3.5.3 Microbial $CaCO_3$ for Self-healing of Cracks in Concrete

When cervices appear, selected bacteria precipitate CaCO3 to heal the concrete cracks. Such self-healing character may lead to the recovery of water tightness and thereby prevention of the entry of corrosive substances into concrete and improves the durability. To apply bacteria for the purpose of self-healing, the bacteria should sustain till the appearance of crack. Due to this reason, spores are recommended in place of vegetative cells for long-lasting results (P. Setlow 1994).

To improve the process of bioprecipitation, nutrients are also added from outside. It was reported that application of yeast extract for *B. sphaericus* speeds up the germination in spores and bioprecipitation at conditions of low temperatures (10 °C) and high concentrations of Ca2+ (Wang 2013). It is very significant from the practical application point of view, as a common suitable temperature range for various bacteria is 20–30 °C, but on the contrary, in many real-life situations, the prevailing low temperature may retard the germination and growth process.

3.6 Control of Biodeteriorating Agents

Currently, various methods are commonly used for the control of this type of biodegradation agents. The treatment of such type of cultural heritage objects against biodeterioration follows preventive and control measures. Preventive measures consist of good housekeeping practice that prevents entry of biodeteriorating agents in such type of object. Proper conditions of storage, prevention of air stagnation in certain areas and sanitation of the premises are important factors to consider in preventing biological attack. Of all the biotic and abiotic factors, humidity, temperature and food are the three most important biological factors which influence the population density of pests thriving on heritage objects. Most of the heritage materials require some kind of chemical protection against biological agencies of deterioration. In some large museums, all new acquisitions are automatically fumigated before being allowed into the collections.

3.6.1 Preventive Methods

These methods are also known as indirect methods. Their main objective is to control and prevent biodeterioration by refining the surroundings where all these museum collections are placed, stored and displayed. In this approach, cultural materials are stored in suitable containers and made airtight so that the exposure to microorganisms, dust, humidity, rodents and insects can be minimized. The purpose of all these activities is to reduce biodeterioration. Along with these works, some other activities are also included like registering and cataloguing of target objects, so that in case of any change in their condition, they can be recognized at initial stage and museums will accomplish documentary control.

3.6.2 Role of Natural Biocides to Prevent Microbial Growth on Cultural Heritage (Biocontrol Methods)

Currently the study of biological function concerning microorganisms led to rapid development in various fields like environmental and medicine biology. Previously, only some microbiological methods particularly traditional ones are used to study the biodeterioration in HC. Nowadays some novel techniques are used for depressing microbial growth (Table 3.1).

During past time, natural mobilization of metals from minerals was supposed to be an abiotic process, and the first acidophilic Fe- and S-oxidizing bacterium was isolated in 1950, from drainage of acid mine, Since from the last century, the potential of microorganisms to encourage the solubilization process of metals has been described, and this type of research studies survived economically also due to commercial interests. The study of acid mine effluents and biomining plants provided us a lot of information about the biodiversity of microbes and ecological interactions involved in mineral weathering derives.

Various historical buildings and artistic objects are made up of stone, etc. Stone is also inevitably deteriorated by various processes, like all other materials. In this inevitable process, different kinds of physical and chemical weathering aspects and microorganism play an important role. Types of stone and local climate play a considerable role in process of biodeterioration including its consequences. Different types of metabolic products are released by various microorganisms such as acids, enzymes, chelating agents, extracellular substances, etc. which are responsible for bio-degradation. At the same time, for biomineralization process, various kinds of heterotrophic and phototropic microorganisms (including actinobacteria and fungus) are the prime damage-causing agents. These microorganisms are able to cause aesthetic damage along with structural injury. In this chapter, it is highlighted that various kinds of isolated bacteria and cell filtrate are able to prevent the growth of the damaging organisms and also admits that treatment of such natural biocides could be an alternative (synthetic biocides) and capable approach to save these cultural heritage monuments, etc. Biological colonization means surface coverage due to growth of microorganisms and original shape is lost, and all these activities are irreversible which leads to transformation of the material. Various kinds of techniques are adopted, direct as well as indirect methods, that permit the management of target object along with suitable management towards surrounding environment.

3.6.3 Biocleaning and Biocidal Methods

3.6.3.1 Organic Matter

While detaching wall paintings, considerable amount of different carbon compounds like casein and glue is disseminated on both surfaces. Such kind of carbon compounds can pose severe threat in process of conversation of artwork. This matter was related to Spinello Aretino, presented in Monumental Cemetery of Pisa. This painting was severely damaged by fire, etc. in 1944. The fresco was removed from the wall, and conservation treatment was applied, and in this treatment, a biocide was also mixed to the glue. This biocide prohibited the growth process of all responsible microorganisms. But with passage of time, polymerization process was favoured, and it posed a danger in removal of such organic substances by conventional methods. In the course of study, a bacteria strain named *P. stutzeri* was chosen as this strain uses various compounds of carbon for its metabolism and is able to manage organic matter. Throughout study period, this microorganism proved to be excellent biorestoration agents on the fresco (Ranalli et al. 2005). The presence of calcium carbonate was considered as major cause for carbonate nature of the plaster on which these paintings were accomplished; gypsum to the plaster sulphatation processes or as consequence of capillary infiltration of walls; and calcium oxalate to the degradation of a superficially applied material.

3.6.3.2 Nitrates

In a research study, nitrate salts were bioremoved with the help of nitrate-reducing bacteria from stony surface of the twelfth-century Matera Cathedral. Exclusion of nitrate was more severe and, even after six years, in areas where bacterial treatment was performed. Concentration of the nitrate continued in stability; no significant presence of colour change or microflora were recorded in comparison with an untreated area. In this concern, the role of physical and chemical treatments did not lead to satisfactory results. Significant reduction up to 92% was recorded when there was use of microorganism (bacteria), viz. *Pseudomonas stutzeri*, when evaluated by ion chromatography. During this study, it has been proved that agar-agar is a suitable delivery system as when it is used to treat vertical surfaces, it sticks very well and leads to homogenous cleaning, and no staining or no residue can be seen on the treated surface.

It was observed that bioprecipitation process of $CaCo_3$ consolidates ornamental surfaces when applied many times, while research studies particularly aiming at identification of microbial species, which may lead to solubilization of carbonate compounds, are still erratic. Furthermore, new supporting material named Laponite was applied, and it was found suitable or conservation resolutions treatments.

During the twentieth and twenty-first centuries, spray paints were frequently used to make paintings, and these contain synthetic polymers. Recent methods which are used to remove these graffiti include physical and chemical including laser techniques. Above all, bioremediation is a great probable technique for graffiti removal. To remove synthetic polymers, different types of microorganisms are used as conservation materials from HC surfaces, e.g. nitrocellulose. Till now only emphasis is laid on the use of graffiti spray paint, which recommends the application of bacterial species. *D. desulfuricans* can be utilized as a potential agent for nitrocellulose degradation.

3.6.4 Routine Maintenance of Buildings

This technique is not as difficult to adopt. It involves the routine care and cleaning of the museum building, etc. It is well known that humidity helps a lot to grow these microorganisms, and uncleaned, dirty surrounding adds to it. Therefore, keeping the concerned building dry, timely repair and maintenance of roofs, etc. to prevent leakage from ventilators, windows and doors. Similarly, cleaning roof, drainage lines and rainspout means that every activity that keeps water, moisture away from the building, should be adopted (Kumar and Kumar 1999).

On the same pattern, outdoors of the museums should be cleaned time to time. It may include activities like cleaning/removal of moss and mould along with algae; still source of dampness of the building can cause major loss, so it should be managed well (Kumar and Kumar 1999).

3.6.5 Housekeeping

It is a primary method for avoiding and controlling the attack of dirt, dust and spores as well as microorganisms. This practice greatly helps in reducing the possibility of bacterial, fungal and other biodeteriorating agents.

3.6.6 Documentation of Collections

The objects of the collection should be written and documented in photographic as well as soft copy records so that their current status will be recorded as well as preventive and corrected action will be managed.

3.7 Factors Influencing Biodeterioration

Many factors influence the activity of microorganisms like humidity, temperature, source of nutrition, pH of the medium, light and pressure. On the surface of ancient art objects, the presence of dust provides a heterogeneous and variable composition. Dust generally provides a nutritive layer for the development of biodeteriorating agents. Microorganisms including fungi and bacteria are carried by visitors to museums and different types of cultural monuments. Type of damage of CH depends on microenvironment of respective site.

3.8 Conclusion and Future Prospects

To prevent such damage, a wide variety of applications and treatments are required. This is the main reason that this process is complex. Treatments can be selected and applied in such a way that there should be no damage to the targeting object. But the most difficult process of this aspect is the environment of the museum. It may be favourable for biodeterioration. Some other problems may also exist, such staff are not trained, because training or devotion to time by museum-related staff to prevent this type of biodeterioration, hence more trained staff for knowledge of conservation methods is needed. Moreover, cost involved and technicalities make the task more difficult, so some other suitable substitute must be found. This chapter includes the evaluation of present applications and treatments, explains various aspects along with preventive and different remedial measures and then scrutinizes precise techniques that are suitable for different kinds of biodeterioration. Two important aspects should be considered before lanuching a biocleaning product in today's market, the delivery system for the microorganism to be used and microorganism itself. Such type of production which aims at biocleaning is also available in market. Nevertheless, the delivery system for microorganisms is one of the major aspects in the bio-cleaning process; Therefore, it was recommended that before choosing these systems, some points must be considered: the ability to deliver microorganisms with precise conditions (such as water, etc.). Application technique should be easy in case of different types of surfaces and irrespective of orientation. Several disadvantages must be considered during conservation treatments in such type of environment, including health problems faced by conservators, besides, further damage to the target object (Webster and May 2006). On the contrary, there are many advantages in the process of biocleaning like no pollution and no effects on the health of conservators. And there should be no residual at the end of application or complete elimination. Finally, we can conclude that biocleaning is a sustainable approach and a greening technique in the preservation of valuable cultural and heritage materials.

After confirmation of the safety and long-lasting effects of these technologies, there is an urgent need for the standardization of the results of respective protocols as well as commercialization of their products used in conservation treatments. From this discussion, we may conclude that further advancement in this field will bring new opportunities to scholars in preservation and renewal of cultural heritage places. addition, nowadays there is an urgent demand for advanced "ready-to-use" type of materials and its production, for the utilization of conservation of our cultural heritage. Ready-to-use, easily available and easy-to-apply biocleaning products include fast treatment and removal of contaminants from the infected objects. At that time proper analysis of cost benefit-ratio biocleaning approach not to only be environmentally justifiable but also economically feasible. Time-to-time surveillance and regular monitoring of developments through this technology over time may provide considerable information, which may answer all significant questions related to the use of such technology and safety.

Under prevailing conditions to cultural heritage property, it is the main duty of conservators and scientists of every country to discuss and exchange views including alternate methods and techniques for conservation of valuable cultural heritage. Even though we have some traditional methods to conserve cultural heritage properties, specific evaluation and parameters, improved advanced techniques to get better and long-lasting results, etc. are not well familiar to us. Considerable importance and proper attention must be given to latest techniques for various activities like inspection, monitoring, etc. concerning interactions among them regarding biocleaned artworks. As far as possible, on-site technologies based on non-invasive tools should be included for better understanding of probable risk along with coverage of physical-chemical, biological and aesthetic analyses. Likewise, future research and complete safety is of prime concern and fundamental importance as this technique has to play very important role in latest technology concerning application of human and eco-friendly sustainable techniques and technologies for conservation of HC properties. Only by spreading awareness about this technique we will be able to preserve HC, as is sometimes miserably occurring currently through the use of conventional toxic organic solvents and violent techniques. To overcome them, intensive, continuous and concerted effort and research are needed.

References

Achal V, Mukherjee A, Goyal S, Reddy M (2012) Corrosion prevention of reinforced concrete with microbial calcite precipitation. ACI Mater J 109:157–164

Anderson S, Appanna VD, Huang J, Viswanatha T (1992) A novel role for calcite in calcium homeostasis. FEBS Lett 308:94–96

Basaran Z (2013) Biomineralization in cement based materials: inoculation of vegetative cells. Ph. D. Thesis, The University of Texas at Austin

Cacchio P, Ercole C, Cappuccio G, Lepidi A (2003) Calcium carbonate precipitation by bacterial strains isolated from a limestone cave and from a loamy soil. Geomicrobiol J 20:85–98

Castanier S, Le Métayer-Levrel G, Perthuisot J-P (1999) Ca-carbonates precipitation and limestone genesis—the microbiogeologist point of view. Sedimentary Geol 126:9–23

De Belie N, Wang J (2016) Bacteria based repair and self-healing of concrete. J Sustain Cement-Based Mater 5:35–56

De Muynck W, De Belie N, Verstraete W (2010) Microbial carbonate precipitation in construction materials: a review. Ecol Eng 36:118–136

Gauri KL, Parks L, Jaynes J, Atlas R (1992) Removal of sulphated-crust from marble using sulphate-reducing bacteria. In: Proceedings of the international conference held in Edinburgh, UK, pp 160–165

Ghosh PS, Mandal BD, Chattopadhyay S, Pal (2005) Use of microorganism to improve the strength of cement mortar. Cem Concr Res 35:1980–1983

Ghosh P, Mandal S, Pal S, Bandyopadhyaya G, Chattopadhyay BD (2006) Development of bioconcrete material using an enrichment culture of novel thermophilic anaerobic bacteria. J Exp Biol 44:336

González-Ruibal A (2005) The need for a decaying past: an archaeology of oblivion in contemporary Galicia (NW Spain). Home Cult 2:129–152

Hideo Arai (2013) Microbiological problems in biodeterioration of cultural property: forty years of study. In: Dhawan S, Abduraheem K, Virendra N (eds) Biodeterioration of cultural Property-7. ICBCP, Sukriti Nikunj, pp 1–10

Horie CV (1987) Materials for conservation: organic consolidants, adhesives and coatings. Butterworths, London

Jroundi F, Fernández-Vivas A, Rodriguez-Navarro C, Bedmar EJ, González-Muñoz MT (2010) Bioconservation of deteriorated monumental calcarenite stone and identification of bacteria with carbonatogenic activity. Microb Ecol 60:39–54

Koestler RJ (2000) When bad things happen to good art. Int Biodet Biodeg 46:259–269

Kumar R, Kumar AV (1999) Biodeterioration of stone in tropical environments: an overview. Getty Publications, Los Angeles

Law JH, Slepecky RA (1961) Assay of poly-β-hydroxybutyric acid. J Bacteriol Res 82:33–36

Le Metayer-Levrel G, Castanier S, Orial G, Loubiere JF, Perthuisot JP (1999) Applications of bacterial carbonatogenesis to the protection and regeneration of limestones in buildings and historic patrimony. Sediment Geol 126:25–34

Lewin SZ, Baer NS (1974) Rationale of the barium hydroxide-urea treatment of decayed stone. Stud Conserv 19:24–35

McNamara CJ, Perry TD, Bearce KA, Hernandez-Duque G, Mitchell R (2006) Epilithic and endolithic bacterial communities in limestone from a Maya archaeological site. Microb Ecol 51:51–64

Orial G, Castanier S, Le Metayer G, Loubière JF (1992) The biomineralization: a new process to protect calcareous stone; applied to historic monuments. In: Biodeterioration of cultural property 2: proceedings of the 2nd international conference on biodeterioration of cultural property, Valladolid, Spain

Park S-J, Park Y-M, Chun W-Y, Kim W-J, Ghim S-Y (2010) Calcite-forming bacteria for compressive strength improvement in mortar. J Microbiol Biotechnol 20:782–788

Price C, Ross K, White G (1988) A further appraisal of the 'lime technique' for limestone consolidation, using a radioactive tracer. Stud Conserv 33:178–186

Ranalli G, Matteini M, Tosini I, Zanardini E, Sorlini C (2000) Bioremediation of cultural heritage: removal of sulphates, nitrates and organic substances. In: Of microbes and art. Springer, Boston, pp 231–245

Ranalli G et al (2005) Biotechnology applied to cultural heritage: biorestoration of frescoes using viable bacterial cells and enzymes. J Appl Microbiol 98:73–83

Rivadeneyra MA, Párraga J, Delgado R, Ramos-Cormenzana A, Delgado G (2004) Biomineralization of carbonates by *Halobacillus trueperi* in solid and liquid media with different salinities. FEMS Microbiol Ecol 48:39–46

Rivadeneyra MA, Martín-Algarra A, Sánchez-Navas A, Martín-Ramos D (2006) Carbonate and phosphate precipitation by Chromohalobacter marismortui. Geomicrobiol J 23:1–13

Rodriguez-Navarro C, Rodriguez-Gallego M, Chekroun KB, Gonzalez-Munoz MT (2003) Conservation of ornamental stone by *Myxococcus xanthus*-induced carbonate biomineralization. Appl Environ Microbiol 69:2182–2193

Sáiz-Jiménez C (1997) Biodeterioration vs biodegradation: the role of microorganisms in the removal of pollutants deposited on historic buildings. Int Biodeterior Biodegradation 40:225–232

Sand W (2001) Microbial corrosion and its inhibition. In: Rehm HJ (ed) Biotechnology, vol 10. Wiley-VCH Verlag GmbH, Weinheim, pp 183–190

Sarda D, Choonia HS, Sarode DD, Lele SS (2009) Biocalcification by *Bacillus pasteurii* urease: a novel application. J Ind Microbiol Biotechnol 36:1111–1115

Setlow P (1994) Mechanisms which contribute to the long–term survival of spores of *Bacillus* species. J Appl Microbiol 76:49–60

Tiano P, Biagiotti L, Mastromei G (1999) Bacterial bio-mediated calcite precipitation for monumental stones conservation: methods of evaluation. J Microbiol Methods 36:139–145

Tiano P, Cantisani E, Sutherland I, Paget J (2006) Biomediated reinforcement of weathered calcareous stones. J Cult Herit 7:49–55

Tilak ST, Pande BN (1997) Biodeteriogens inside a library. In: Agashe SN (ed) Aerobiology. Pub. Oxford & IBH, New Delhi, pp 177–189

Verma RK, Chaurasia L, Bisht V, Thakur M (2015) Bio-mineralization and bacterial carbonate precipitation in mortar and concrete. J Biosci Bioeng India 1:5–11

Wang J (2013) Self-healing concrete by means of immobilized carbonate precipitating bacteria. PhD thesis, Ghent University, Belgium

Wang J, Soens H, Verstraete W, De Belie N (2014) Self-healing concrete by use of microencapsulated bacterial spores. Cem Concr Res 56:139–152

Warscheid T, Braams J (2000) Biodeterioration of stone: a review. Int Biodeterior Biodegradation 46:343–368

Webster A, May E (2006) Bioremediation of weathered-building stone surfaces. Trends Biotechnol 24:255–260

Entomogenous Fungi and the Conservation of the Cultural Heritage

4

Saritha Vara, Manoj Kumar Karnena, Swathi Dash, and R. Sanjana

Abstract

Entomogenous fungi are agents of deterioration of cultural heritage. Their capabilities to survive at low water activities, thrive in microclimatic niches, etc. enable them to adapt to various conditions and colonize on cultural heritage leading to their deterioration. An understanding of the properties of these organisms will enable to propose countermeasures for protection of heritage. Material-specific cleaning and application of precise biocides are necessary for the conservation of cultural heritage. The key focus of this chapter is to apprehend the importance of cultural heritage, threats to cultural heritage and role and mechanism of entomogenous fungi in their deterioration. Moreover, an insight into conservation methods from traditional towards modern is presented with special emphasis on contribution of nanoscience for conservation of all types of cultural heritage. Probable directions for unravelling numerous conservation issues that still need to be faced in the future are also emphasized.

Keywords

Entomogenous fungi · Cultural heritage · Conservation · Nanoscience

4.1 Introduction

Cultural heritage is understood as cultural inheritance corresponding to any given community which is protected in its inherent condition and has been transferred to present as well as will be transferred to future generations. Being subjective and dynamic, cultural heritage is not dependent on substances or possessions but is

S. Vara (✉) · M. K. Karnena · S. Dash · R. Sanjana
Department of Environmental Science, GITAM Institute of Science, GITAM (Deemed to be University), Visakhapatnam, Andhra Pradesh, India

A. N. Yadav et al. (eds.), *Microbial Biotechnology Approaches to Monuments of Cultural Heritage*, https://doi.org/10.1007/978-981-15-3401-0_4

dependent on ethics and values of society which are generally attributed to them at the respective moment of history, determining the goods which are to be protected and preserved for generations to come. Further, it is understood that both the thinking process and behaviour of a community are strongly dependent on the legacy of physical artefacts and intangible attributes inherited from past generations, carried on in the present and possibly bestowed to the benefit of next generations to come. Hence, identification, fortification, preservation and propagation of world's cultural heritage are among the most recognized responsibilities of United Nations Organization for Education, Science and Culture (UNESCO 2013; Cuetos 2012).

Cultural heritage constitutes both tangible and intangible assets that are left by history to a country and its citizens. It has exceptional and relevant scientific, historical, symbolic and aesthetic importance. Such inheritance received from our ancestors today stands testimony of their existence and visualization of the biosphere (Rivera 2018). Tangible cultural heritage, also known as material heritage, comprises of property both movable and immovable such as art collections of ethnographic, religious, historical, technological, archaeological, artistic and artisan objects. It also includes physical sites of archaeology, buildings, engineering works and ensembles of architecture. Intangible cultural heritage encompasses treasure of knowledge and expressions inherited from our ancestors communicated to next generations like oral traditions, language, techniques, performing arts, traditional crafts, customs, rituals, ways of life, festivals and practices in relation to nature and universe (ICOMOS 2002). In a nutshell, cultural heritage is understood to be man's legacy which will be carried on to future generations.

Both tangible and intangible heritages possess vital associations between past and present, playing a critical role in shaping the present communities. Further, rational and well-accomplished intellectual and physical access to cultural heritage is both a privilege and a right. Cultural heritage receives a duty of respect, values, interests and equity from present-day community hosting it. Hence, appraising knowledge repository related to cultural heritage is mandatory for preservation, conservation and appropriate management.

The objective of preservation of cultural heritage comprises four principal reasons. First, it is understood as a cultural memory which sustains physical evidence of history and transfers intangible resource like knowledge and skill sets of our ancestors. Second, deemed to be convenient proximity, it supports the interactions among people, environment and activities of the community. Third, diversity of environment cultural heritage is personality of the local community, and preservation would retain inherent artefacts and artisans among the changing urban conditions. Finally, monetary gain has added benefit to the community which will not only save the cost of new structures but also attracts visitors (Prompayuk and Chairattananon 2016).

4.2 Impendence to Cultural Heritage

Cultural heritage can be either movable or immovable objects. The movable group comprises of historical and documentary books and manuscripts made up of parchment or paper, easel paintings on canvas or wood and a wide variety of objects like statuettes, textiles and jewellery. The group of immovable cultural heritage comprises of wall/mural paintings, substrates of architectonic, statues and several other kinds of stone-based artefacts. Irrespective of their nature, artefacts are irreparably suspected to various degradation factors like light, relative humidity, temperature, physical erosion, chemical degradation and microorganisms leading to natural ageing of the heritage. Furthermore, anthropogenic activities releasing higher concentrations of atmospheric pollutants like oxides of sulphur, nitrogen and volatile organic compounds eventually lead to corrosion of heritage.

Intensified social and cultural changes demand greater conservative measures in order to restrain unwanted changes. The necessity for conservation of cultural heritage does not only emanate as significant evidence of past history but also helps in forging a solid connection between past and present in the memory of the community. Preserved as a legacy, cultural heritage is steadily being eroded and threatened due to insensitive urbanization and modernization. This calls for immediate attention towards conservation practices and precise understanding of various causes, factors affecting cultural heritage. Conservation of cultural heritage is challenging owing to factors influencing their conditions compromising storage conditions, climate change or adversities, leading to deterioration or total loss. Risks to cultural heritage vary from impulsive and calamitous events like earthquakes, floods and fires to steady and cumulative process like chemical, physical and biological degradation. Understanding mechanisms governing these processes along with their effects on changing cultural heritage will enable rational use of these besides anticipating their changing trends beforehand towards succeeding in conservation and eventual restoration.

Degradation of cultural heritage is a process that depends on external agents, exposure, inherent properties of materials, etc. With severe and frequent weather events and longer duration of exposure leading to enhanced degradation, the need for new conservation methods in line with understanding the root cause has become a need of immediate attention (Bertolin 2019). Biodeterioration of organic material is a vital process in nature enabling recycling of composite organic material which is an integral component of life. However, this process destroys historical records which are composed of organic matter precisely natural and synthetic polymers, leading to loss of valuably treasured cultural heritage (Cheng and Gross 2003). The continuous cycle of disintegration and reconstruction results in loss of original physical, chemical and optical qualities of the materials. The process of disintegration and biodeterioration is aggravated by two types of agents (Vásquez Ponce 2013; Simmons and Muñoz-Saba 2005):

1. Intrinsic: depends on the nature of material, manufacturing technique and procedures used to perform the work
2. Extrinsic: depends on external sources like factors concerning the environment, anthropogenic, biological and catastrophic

The three key mechanisms of deterioration are physical or mechanical processes, changing the behaviour of the material; chemical processes, transforming the material; and biological processes where insects and microbes bring about the deterioration (Hueck 1965). In general fungi are understood to be recurrent and harmful microorganisms that are associated with biodeterioration of both inorganic and organic materials; the high metabolic versatility of fungi enables them to colonize a wide variety of substrates including stone, wood, paper, etc. leading to initial and destructive steps like damage of aesthetic and mechanical or chemical nature (Sileo et al. 2015; Ruga et al. 2015). Thus continuous analysis, monitoring, interpretation and understanding of environmental parameters giving rise to different types of degradation, precisely those of biological origin, are mandatory (Ruga et al. 2008).

The role of microorganisms in the deterioration of cultural artefacts is a well-established fact. Two key factors that favour microbial growth on cultural heritage are chemical composition and nature of the substratum and environmental conditions (Saiz-Jimenez 1993). Paintings are composed of both inorganic and organic compounds which are taken by majority of microorganisms for their growth and proliferation. Some of these include cellulose of the painting canvas, animal glue and gypsum utilized in the preparation of the ground layer and linseed oil of paint layer. Further, spectrum of compounds providing nutrients to microbes is augmented from dust, dirt and other environmental contaminants deposited on painting surface along with dead or living cells that accumulate on the painted surface which provide additional source of nutrients (Ciferri 1993; Wainwright et al. 1993; Flemming 1998; Warscheid 2000).

One of the major causative factors for degradation of cultural heritage is biofilms formed by mono- or multi-species (consortium) of microorganisms. Biofilms are considered as microbial communities growing in embedded self-produced exopolysaccharide matrix attached to an inert surface or living tissue (Costerton et al. 1987; Estela and Alejandro 2012; Castrillón et al. 2013). The requirements for the formation of biofilm are surface, nutrients, moisture and microorganisms. Being a complex microbial organization, biofilms provide several benefits for their survival like resistance to stress from environment by formation of stable microcolonies, facilitating genetic material exchange and accumulation of water and nutrients in its matrix offering defence against toxic substances like antibiotics and biocides (Rivera et al. 2010).

The role of biofilms in biodeterioration of cultural heritage has been reported for several decades owing to their relation to:

(a) pH and ionic concentration modifications
(b) Reduced oxide concentrations in the interface of substrate and biofilms

(c) Masking properties and covering a surface of biofilms
(d) Enhancing additive and monomer leaching outside the polymer matrix due to biodegradation
(e) Release of enzymes causing loss of mechanical stability or embrittlement
(f) Accumulating water penetrating the matrix leading to swelling and enhanced conductivity
(g) Excretion of lipophilic pigments (Morton and Surman 1994; Videla et al. 2003)

4.3 Fungal Mechanisms in Deterioration of Cultural Heritage (Mycodeterioration)

The association of cultural heritage and mould comes from a long history, including myths and mysteries, whether it is the "curse of the pharaoh" – death of several archaeologists after opening of Tutankhamun's tomb which was later explained as the death may be possibly due to the fact of the presence of spores of pathogenic fungi *Aspergillus niger* and *Aspergillus flavus* that infected the lung or caused other systemic mycosis called aspergillosis.

Fungi were always and still are threats to cultural heritage and also contemporary objects of art (Sterflinger 2010; Pangallo et al. 2009; Mesquita et al. 2009; Cappitelli et al. 2009; Allsopp et al., 2004; Koestler and Koestler 2003; Nittérus 2000). Fungi are also understood to be existing on caves, mural paintings and catacombs and on the surface of architectural and stone monuments (Steiger et al. 2011; Ettenauer et al. 2012; Ettenauer et al. 2010; Saarela et al. 2004; Sterflinger 2000). The most precious old cultural heritage suffering from serious fungal invasions are rock art caves of Lascaux of France (Bastian and Alabouvette 2009).

Based on phylogenetic studies, fungal kingdom is currently subdivided into five valid divisions (James et al. 2006), *Chytridiomycota*, *Glomeromycota*, *Zygomycota*, *Basidiomycota* and *Ascomycota*. Among these, *Basidiomycota* comprises of important wood-decaying fungi, damaging indoor wood and historical buildings, e.g. *Serpula lacrymans* (Bech-Andersen et al. 1993). *Ascomycota* comprise of most of the mould species that play a key role in deterioration of cultural heritage.

On the basis of cultural heritage, biodegradation fungi are subdivided into two key functional groups: opportunistic fungi growing on practically all types of materials in the presence of sufficient humidity. These cannot degrade the material enzymatically (**entomopathogenic fungi**). Material pathogens are substrate-specific fungi which are able to degrade precisely substrate-specific materials, e.g. cellulolytic fungi on paper and keratinolytic fungi on leather, feather and hair (Błyskal 2009; Meier and Petersen 2006). Though both groups cause deterioration, the second group causes decay of the material itself.

Table 4.1 Fungal genus with some entomogenous species

Genera	References
Aspergillus	Leatherdale (1970), Humber and Hansen (2005) and Greif and Currah (2007)
Cladosporium	Leatherdale (1970), Gunde-Cimerman et al.(1998), Humber and Hansen (2005) and Greif and Currah (2007)
Fusarium	Leatherdale (1970), Teetor-Barsch and Roberts (1983), Claydon (1984) and Humber and Hansen (2005)
Geotrichum	Gunde-Cimerman et al. (1998) and Humber and Hansen (2005)
Gliocladium	Gunde-Cimerman et al. (1998), Humber and Hansen (2005) and Greif and Currah (2007)
Mucor	Gunde-Cimerman et al. (1998), Humber and Hansen (2005) and Greif and Currah (2007)
Penicillium	Gunde-Cimerman et al. (1998), Humber and Hansen (2005) and Greif and Currah (2007)
Phialophora	Leatherdale (1970), Humber and Hansen (2005) and Greif and Currah (2007)
Phoma	Gunde-Cimerman et al. (1998) and Humber and Hansen (2005)
Scopulariopsis	Gunde-Cimerman et al. (1998), Humber and Hansen (2005) and Greif and Currah (2007)
Stachybotrys	Humber and Hansen (2005)
Trichoderma	Humber and Hansen (2005)
Trichothecium	Humber and Hansen (2005)

Source: Jurado et al. (2008)

4.3.1 Entomopathogenic Fungi

Belonging to 12 classes among the 6 phyla of the kingdom *Fungi*, entomopathogenic fungi are pathogenic to arthropods found in divisions of *Deuteromycota*, *Ascomycota*, *Oomycota*, *Chytridiomycota* and *Zygomycota* (Maina et al. 2018; Saiz-Jimenez and Samson 1981; Shahid et al. 2012) (Table 4.1).

Sources of fungi in any environment are owed to the hydrophobic spores and conidia which are easily transported by wind; hence fungal deterioration of cultural heritage is mostly airborne, varying seasonally with a large number of spores accumulated in dust layers (Kaarakainen et al. 2009). Precisely entomopathogenic fungi find their way from the infected animals in and around the heritage zones. Spore germination leading to colony development is governed by factors like chemical composition of material, environmental factors, airborne nutrients and availability of water – aw above 0.8 proliferates growth of a wide variety of airborne fungi at temperatures ranging between 15 and 35 °C. Further, key enzymes produced by entomopathogenic fungi which assist in biodegradation include lipases, proteases, chitinases, β-galactosidase, catalase, L-glutaminase.

4.4 Mechanisms Comprising Mycodeterioration of Paintings

4.4.1 Mural Paintings

Paintings are critical constituents of cultural heritage around the world. These contain organic and inorganic nutrients which stimulate microbial growth. Most of the components in paint and paintings (easel or mural) are biodegradable which include constituents like sugars, gums, waxes, oils, proteins and polysaccharides and other components like glues, emulsifier, thickener, etc. (Gettens, et al. 1941; Ciferri 1999) which provide varied ecological niches to be exploited by a wide variety of microbes. In presence of favourable conditions like relative humidity and temperature, the microbes attack the paint surface leading to changes in structure, discolouration of pigment, paint blister formation, cracking and paint layer detachment (Petrov et al. 2007). Degradation takes place through the extracellular enzymes produced by fungi which include ligninases, celluloses and organic acids which facilitate them to colonize almost all substrates which are capable of supplying free carbon as source of nutrient. Further, environmental pollutants like soot and dirt accumulated on painted surface are other trace contributors of nutrients.

Key mechanisms of painting biodeterioration include pigmentation, penetration or hydration into the materials and degradation of compounds, leading to colonization of paintings excreting aggressive metabolic products along with producing extracellular enzymes that further deteriorate the material. Chief enzymatic activities that carry out deterioration are proteases, lipases and esterases. Hydrolysis of carboxyl ester bonds is catalysed by lipases and esterases (Chahinian et al. 2002). Furthermore, enzymes playing important role in deterioration are endo-N-acetyl-P-D-glucosaminidases (ENGases), hydrolysing glycosidic bond between N-acetyl-P-D-glucosamine residue and adjacent monosaccharide within oligosaccharide chain. Among the three types of ENGases, type II ENGases which act on chitin are the chief components of cell walls of fungi (Karamanos 1997). Few chitinases are known to display somewhat pronounced lysozyme activity. Fungal ability to grow at lower water activity values (aw) produces large amounts of exoenzymes like glucanases, cellulases, phenolases, laccases, keratinases, mono-oxygenases and many others which affect the oil paintings (Sterflinger, 2010).

Several studies have reported a fungal deterioration of paintings (Guglielminetti et al. 1994; Nugari et al. 1993; Inoue and Koyano 1991; Agrawal et al. 1988; Arai 1984; Giacobini and Firpi 1981; Saiz-Jimenez and Samson 1981; Tiano, and Gargani 1981; Savulescu and Ionita 1971; Gargani 1968; Obidi et al. 2009). Precisely more than 33 dissimilar fungal species belonging to 17 genera were isolated from fresco in St. Damian's Monastery in Assisi, Italy (Guglielminetti et al. 1994). *Beauveria alba*, an entomopathogenic fungus, was repeatedly isolated from mural painting of Canterbury Cathedral (Jeffries 1986). Little more than 10% of *Engyodontium album* were isolated from black and green lesions. Fungi are considered as secondary colonizers of fresco, first being sulphur-cycling bacteria. Death and decay of bacteria provide organic substrates required for growth of heterotrophic fungi (Saiz-Jimenez and Samson 1981; Bock 1993) which are

Table 4.2 Most representative entomogenous genera of fungi in cave and mural paintings

Genera	References
Acremonium	Ionita (1973), Saiz-Jimenez and Samson (1981), Agrawal et al. (1988), Berner et al. (1997) and Gorbushina and Petersen (2000)
Beauveria	Mason-Williams (1965), Gorbushina and Petersen (2000) and Stankovič et al. (2005)
Cordyceps	Matočec and Ozimec (2001) and Kubátová and Dvorák (2005)
Engyodontium	Saiz-Jimenez and Samson (1981), Jeffries (1986), Berner et al. (1997) and Gorbushina and Petersen (2000)
Entomophaga	Gunde-Cimerman et al. (1998)
Entomophthora	Kubátová and Dvorák (2005)
Hirsutella	Mason-Williams (1965)
Lecanicillium	Saarela et al. (2004)
Metarhizium	Gunde-Cimerman et al. (1998)
Isaria	Kubátová and Dvorák (2005) and Dupont et al. (2007)
Verticillium	Went (1969), Berner et al. (1997), Gorbushina and Petersen (2000) and Dupont et al. (2007)
Torrubiella	Saarela et al. (2004)

responsible for stains on the fresco's surface and mechanical damage. Thus, it is understood that environmental pollutants like oxides of sulphur cause direct damage to fresco and provide substrate required to promote growth of aerobic and anaerobic sulphate-cycling bacteria which supply organic nutrients that enable establishment of scavenger fungal community which further degrade fresco (Table 4.2).

4.4.2 Canvas Paintings

The key component of paintings on canvas is pigmented, which are either natural or synthetic in origin with three chief functions like providing colour, brilliance and opacity and protecting the surface on which they are applied along with protecting binder from destruction by UV radiation. Adhesives are basic elements to achieve final result of the painting which facilitate uniform colour distribution, preventing the paint layer from being absorbed by fabric. Some examples of the adhesives which have changed over the years are either from animal (wax, gelatin, casein and albumin) or plant (gluten, resins, starch and gums) origin. Biodegradation of canvas is dependent on factors like composition of fabric formed by cellulose fibres. Degradation of cellulose involves various enzymes which act together to obtain glucose molecules which can be assimilated by microorganisms for their carbon source. The vulnerability to biological degradation is dependent on content of lignin and cellulose along with other organic components. The cellulolytic degradation process is progressive in conditions with high relative humidity. Other constituents that are vulnerable to biodegradation are starches, gums, adhesives, varnishes, resins, etc. all of which are natural in origin (Ravikumar et al. 2012; López-Miras et al. 2013; Sterflinger 2010; Mesquita et al. 2009; Domenech-Carbo 2008; Pankhurst, et al., 1972; Nugari and Priori 1985).

Canvas painting biodeterioration by microbial agents are formed first by alterations of canvas carried out by loss of strength or support, hydrolysis, cracks, colourations, deformations and scales; second, by alternations of adhesive through disintegration, enzymatic degradation, pulverulence and colourations; and finally by alterations of varnish through peeling, yellowing, whitening and tiling (Poyatos Jiménez 2007). Entomopathogenic fungi that are commonly found on oil paintings are *Aspergillus*, *Fusarium*, *Penicillium*, *Aureobasidium* and *Trichoderma* (Pangallo et al. 2009; Inoue and Koyano 1991; Ciferri et al. 1996). It was reported that *Aureobasidium pullulans* were the most frequent isolate from oil painting that uses cellulose produced from bacteria for growth (Schmitt 1974). Fungal species that deteriorate the precise components of the paint are *Penicillium* and *Aspergillus* disintegrating oil binders, while *Aureobasidium* decompose oil binders (Geweely 2006). Further, Dhawan and Agrawal (1986) reported that cellulolytic content is the one that is hydrolysed first by fungal species, and hence, those fungi with superior cellulase yield are more prominent on canvas paintings, and the cellulolytic activity is dependent on mechanism of induction and catabolite repression of individual species. The cellulolytic fungi include *Aspergillus*, *Penicillium* and *Trichoderma.*

4.5 Mechanisms Comprising Mycodeterioration of Textual Heritage

4.5.1 Papyrus

Papyrus composes of cellulose and natural sugars, which make it susceptible to fungal degradation. Though many fungi are involved in the degradation of papyrus, the literature on entomopathogenic fungal degradation of papyrus is not available. In general, utilization of inorganic nitrogen source by fungi was in the order of ammonium nitrate > ammonium phosphate > ammonium sulphate > sodium nitrate, showing that ammonium nitrate is a good source of nitrogen. Further, most of the fungi degrade through hydrolysis of monosaccharides, and it has been reported that L-rhamnose and DL-arabinose were more certainly degradable than D-arabinose (Menei 1990) (Table 4.3).

4.5.2 Parchment

Composition of parchment includes collagen, and its degradation is dependent on oxidative chemical deterioration of amino acid chains, hydrolytic breakdown of peptide structure and production of pigments and organic acids which modify parchment. Probable entomopathogenic fungal genera identified in degradation of parchment using proteolytic enzymes include *Acremonium*, *Fusarium*, *Aspergillus*,

Table 4.3 Entomopathogenic fungi with corresponding enzymes

Enzymes	Entomopathogenic fungi
Lipase	*Fusarium oxysporum, Metarhizium anisopliae, Aspergillus flavus, Beauveria bassiana*
Protease	*Metarhizium anisopliae, Beauveria bassiana, Verticillium lecanii, Paecilomyces fumosoroseus, Isaria fumosorosea, Tolypocladium niveum*
Chitinase	*Trichoderma atroviride, Trichoderma harzianum, Trichoderma virens, Metarhizium anisopliae, Beauveria bassiana, Nomuraea rileyi, Aschersonia aleyrodis, Verticillium lecanii, Isaria fumosorosea*
β-Galactosidase	*Aspergillus* spp., *Aspergillus foelidis, Beauveria bassiana, Aspergillus fonsecaeus, Aspergillus oryzae, Aureobasidium pullulans, Curvularia inaequalis, Fusarium moniliforme, Metarhizium anisopliae, Metarhizium robertsii*
Catalase	*Lecanicillium muscarium, Fusarium oxysporum, Verticillium dahliae, Aspergillus phoenicis*
L-Glutaminase	*Beauveria bassiana, Trichoderma koningii, Aspergillus flavus, Acremonium furcatum, Aspergillus wentii MTCC 1901, Trichoderma harzianum*

Source: Subhoshmita Mondal et al. (2016)

Penicillium, *Verticillium* and *Trichoderma* (Sterflinger 2010, Petushkova and Lyalikova 1986; Agarwal and Puvathingal 1969).

Susceptibility of parchment to microbial degradation is governed by factors like raw material, production methods and preservation conditions. Studies have shown that biodeterioration of various types of leather at high relative humidity resulted in the development of various fungi which attacked tannins, collagen fibres and water-insoluble substances (Strzelczyk et al. 1989).

4.5.3 Paper

With the invention of paper, man has found faithful support with written memory which can record his journey through history, and hence, the texts of our ancestors are still, today, faithful witness of their time. Most of our tangible cultural heritage is treasured in paper objects ranging from books, documents, maps, paintings, drawings and other artworks (Roman et al. 2013). The organic composition of paper-based heritage makes it highly susceptible to irreversible biodegradation, whose protection is of great significance (Melo et al. 2019; Allsopp et al. 2004; Area and Ceradame 2011; Corte et al. 2003).

The key components of the paper include fibre from varied materials like cotton, bagasse, linen, hemp, wood and rice straw and functional additives like optical brighteners and sizing and consolidating agents like cellulose acetate, gelatin and carboxymethyl cellulose. Another important component of paper-based heritage is ink consisting of a liquid fixed to the support. The ink is composed of pigment, diluent and binder. The oldest inks are ferrous inks composed of gallotannic acid, iron sulphate and binder – gum arabic (Vaillant 2013).

Fungi are understood to be among the most degrading microorganisms of cultural heritage (Coutinho et al. 2019); in particular, those fungi secreting cellulolytic enzymes (*Beauveria Bassiana*) are supposed to be more harmful to paper-based heritage (Sterflinger 2010; Fabbri et al. 1997). Fungal effects on paper heritage are discolouration, damage of structural integrity, etc. (Strzelczyk 2004; Arai 2000). The key mechanism of paper heritage degradation is through cellulose degradation (Jerusik 2010; Cappitelli and Sorlini 2005; Kowalik 1980), wherein acids are formed from oxidation of glucopyranose rings and hydrolysis of glycosidic bonds resulting in aggravated catalysis of paper by breaking cellulose chains leading to transformation of paper to brittle and delicate object. Further, the damage varies from erosion and formation of age spots which might be more or less pronounced (Florian 2002; Gallo 1965). Two of the most prolific cellulose producers of fungi threating paper heritage are from genera *Trichoderma* (entomopathogenic fungal species capable of secreting cellulases – *Trichoderma atroviride*, *Trichoderma harzianum*, *Trichoderma virens*) and *Aspergillus* (*Aspergillus oryzae*) (Pinzari and Montanari 2011; de Vries and Visser 2001; Domingues et al. 2013). Factors governing or aggravating the biological attack include environmental relative humidity and associated water absorption levels (Nielsen et al. 2004). Greater than 10% of water content enhances germination and growth of certain species (Hoang et al. 2010). The demand for water is dependent on species which can be either xerophilous or hydrophilous (Nyuksha 1983; Ciferri et al. 2000; Gallo 1965)

4.6 Mechanisms Comprising Mycodeterioration of Fabric

4.6.1 Textiles

Considered as representatives of cultural identity, textiles possess a significant value that transcends their material and work needed for their manufacture. All cultures express and communicate their social, cultural and aesthetic values through their material and textile manifestations like clothing which are unique cultural heritage (Dakal and Arora 2012). Being very extensive, textile heritage is prone to be lost throughout history, owing to its characteristic of delicate materials. Gradual deterioration of this material is sensitive owing to inadequate handling and bad storage conditions leading to unrepeatable loss of textiles since these materials and their preparation process are from our past (Olmedo-Juárez et al. 2017).

Mycodegradation of these materials is through the process of assimilation by fungi which use these materials as a source of nutrients for their metabolism; further this process is aggravated by the presence of light, humidity and temperature. Key manifestations of mycoremediation include changes in the surface of material, pH, reduction in material resistance, discolouration and unpleasant odour. Effects of mycodeterioration of textiles include total destruction of material, reduction of degree of polymerization, tensile strength and elasticity. Exopigments of fungi are considered to be key molecules responsible for mycodeterioration of textile fibres from the species of *Penicillium* and *Aspergillus* (Joshi and Attri 2005). Mechanism of textile mycodeterioration is dependent on types of fibre: cellulose fibres, wool fibres or silk fibres.

4.6.2 Cellulose Fibres

Degradation of cellulose or (1,4)-β-D-glucan is due to the activity of cellulolytic enzymes produced by fungi which hydrolyse cellulose to glucose. Mycoremediation of cellulose fibres is governed by physical and chemical properties of fibres (Tomšič et al. 2011; Valentin 2003) facilitated by improper storage conditions. Fungi are understood to be the most severe degraders of cellulose fibres (Seves et al. 1998; Szostak-Kotowa 2004). These enzymes reduce the degree of polymerization of long-chain cellulose molecules resulting in reduced strength of the fibre. Presence of other fibre components like pectins, hemicellulose and other carbohydrates provide additional nutrients to the fungi. Major fungal genera associated with mycodeterioration are *Aspergillus*, *Penicillium*, *Trichoderma*, *Fusarium*, *Verticillium* and *Paecilomyces* (Pekhtasheva et al. 2012).

4.6.3 Wool Fibres

A chief component of wool fibres is keratin, a protein that forms polymer when disulphide bridges cross over. Mechanism of mycodeterioration of wool fibres is carried out by proteolysis by peptidases, sulfitolysis, keratinolysis and deamination. Fungal growth and metabolic activity result in structural decomposition of fabric leading to a decrease in textile strength. The extent of fungal growth and destruction of wool fibres is regulated by physiological potential of the fungi enabling use of nutrients from the material on which the organism grows and environmental factors (Błyskal 2005, 2009, 2014, 2015). The magnitude of degradation is dependent on molecular structure, chemical composition, degree of polymerization of substrate and structure of keratin (Kunert 1989). Key fungi causing cellulose mycodeterioration are *Fusarium*, *Aspergillus*, *Acremonium*, *Penicillium* and *Trichoderma* (Agarwal and Puvathingal 1969).

4.6.4 Silk Fibres

Produced by silkworms, silk fibres are fibroin proteins (Vollrath 1999; Seves et al. 1998) which are united to each other by rubber-like proteins called sericin (Szostak-Kotowa 2004) which assist as shield from damage by light. Silk fibres possess high resistance to mycoremediation, whose decomposition is dependent on proteolytic action on sericin and fibroin which are taken up as carbon source by fungi like *Penicillium*, *Aspergillus* and *Rhizopus* (Gutarowska et al. 2017; Ciferri et al. 1996; Abdel-Kareem 2010). The results of mycodeterioration mechanical strength of silk fibres is lost making them brittle (Milicia 2014).

4.7 Mechanisms Comprising Mycodeterioration of Stone

Surface properties and mineralogical nature of stone along with environmental conditions synergistically act for its bioreceptivity. Further, intensity of mycodeterioration is dependent on pollutant concentration, manmade atmospheric eutrophication and microclimatic conditions (Dakal and Cameotra 2012). Climatic conditions affecting the stone include wind which wears the rock resulting in erosion, solar radiation leading to discolouration, temperature, rain and humidity that induce physical and chemical wear and erosion. All these factors affect stability of matrix through chemical corrosion forming minerals through oxidation, hydration and dissolution of carbonates and solubilization of few elements of mineral silicates (Tiano 2016; Castro and López 2013). Microbial communities upon interaction with biotic and abiotic factors develop using stone as substrate partially being responsible for chemical and physical deterioration of stone altering the aesthetic appearance and physical integrity of material through various mechanisms.

Effects of mycodeterioration on stone include water retention, discolouration, growth stimulation of heterotrophic and higher organisms, disintegration and breakage of material, patina formation, corrosion, alkaline dissolution and alteration of stratified silicates (Cameotra and Dakal 2012). Colonization of microbes on surface of bare stone is believed to be instigated by pioneering microbes like lithophiles, photoautotrophs and chemolithotrophs, which might secrete carbohydrates and growth factors helping formation of a biofilm thus supporting growth of successive microbial communities (Dakal and Cameotra 2012).

Exopigments produced by fungi with different colours are black, blue, purple, green and violet. Black pigment is known as melanin that protects fungi against cellular lysis and environmental threats. Further, carotenoids and mycosporines protect fungi against extreme UV radiation and act as antioxidants and osmoprotectants along with providing tolerance against desiccation (Salvadori and Municchia 2016; López-Miras et al. 2013). The another mode of deterioration of stone by fungi is mechanical where physical intrusion and fungal hyphae penetration in gaps, pores and fractures is the phenomenon. This intrusion destabilizes the texture of stone leading to mechanical deterioration.

Wear of rocks is also believed to be a result of removal and solubilization of cations in the minerals of stone precisely iron and manganese by negatively charged exopolysaccharides of biofilms and siderophores (microbial proteins) by complexes of organic transport and metallic organic chelates. During conditions of low iron stress, iron is chelated by siderophores and supplied to fungal cells through receptors of outer membrane. Siderophores scavenge iron from environment and make it available to microbial cells (Neilands 1995). Wearing of stone is aggravated due to the presence of biofilms which contain latent and active microorganisms along with their metabolic products like organic and inorganic corrosive acids (oxalic, lactic and gluconic acids) (Petersen et al. 1988; Di Bonaventura et al. 1999) and polymeric materials, which act as gums trapping dust and particulate matter thus increasing the disfiguring effects of biofilm (Gaylarde and Morton 1999).

Owing to the complex metabolic activities on stone surfaces, fungi play a dangerous role in their biodeterioration. Some of the entomopathogenic fungal species which are prevalent on stones that were isolated from monuments around the world are *Acremonium*, *Aspergillus*, *Aureobasidium*, *Fusarium*, *Penicillium* and *Trichoderma* (Rebricova 1991; Xie 1991; Gorbushina et al. 2004; Strzelczyk 2004). Studies on isolation of fungal strains and their organic acids from archaeological monuments of Dharmarajika have clearly indicated the mechanisms of patinas and stains, which was correlated with different coloured hard stains (oxalate films) on limestone-based monuments of Taxila. Yellow, red and brown patinas were studied on marble buildings of Italy which has the chemical composition of calcium oxalate (Monte 2003).

4.8 Conservation of Cultural Heritage

A multifaceted discipline dealing with restoration and conservation of a wide variety of materials of cultural heritage is called conservation science. Conservation is a process of continuous representation of phenomenon related to cultural heritage which is constantly in flux from local to global scale. Origin of modern conservation science dates back to 1966 when tragic floods have devastated Florence and Venice, which has imposed a need to search for new methodologies for restoration and conservation of vast cultural heritage. Enzo Ferroni was the first to propose a scientific method towards conservation of carbotic materials (Ferroni et al. 1969; Ferroni and Baglioni 1984). An "in situ consolidation" method for wall paintings was successful in restoration through two-step processes of chemical reactions which include degradation of calcium carbonate that is the chief constituent in wall paintings.

Modern conservation science that has developed rapidly yet in a healthy way had two main streams (Baglioni et al. 2012):

1. Characterization of works of art materials analytically, characterization of pictorial techniques adopted by artists and chemical reactions leading to their degradation
2. Search for advanced scientific methods towards conservation and restoration of our cultural heritage so that they can be transferred to the next generation

Conservation of tangible cultural heritage consists of two ways of action:

1. Prevention of deterioration called conservation
2. Repair of damage known as restoration

Preventive measures are those which are not directly applied to the object but instead are directed to the environment controlling microclimatic conditions in order to eradicate harmful agents either temporarily or permanently (Pinzari and Montanari 2011). On the other hand, conservation is defined as a set of operations aiming to prolong life of the material. Conventional methods for control of biodeterioration are mechanical, physical and biological.

4.8.1 Mechanical Methods

These remove fungi physically by abrasion, shaving, brushing, etc. Results obtained are short term, but this method is not suitable for long-term conservation, and also complete elimination is not attained.

4.8.1.1 Physical Methods

These include modification in environmental conditions like temperature and pressure which will not allow the growth of organisms. Further, the biocidal effect of the organism is reduced by denaturing the molecules of the organism by treating with electromagnetic radiation, extreme temperatures and anoxic treatments. These methods remove microbes by directly attacking their genetic material, through alteration of their structure and metabolic function. Drawbacks of these methods are due to their cost and probability of damage to materials by alerting their chemical nature resulting in pigmentation and hydrolysis of cellulose and proteins.

4.8.1.2 Chemical Methods

These are the most frequently used intervention methods through biocides like fungicides or general disinfectants. They are applied in gaseous or liquid form, their mechanism of action varies with the chemical, and they attack through disintegrating fungal membrane or inhibiting cellular processes leading to death upon application of appropriate dose. Their use has been limited owing to high risk and limited knowledge with reference to compatibility with the material of application (Melo et al. 2019). A good biocide is one which possesses a broad spectrum of activity, is effective at lower concentrations and effective over a wide range of pH, has high persistence and moreover has low human and environmental toxicity with low cost (Borrego-Alonso 2015).

4.8.1.3 Biological Methods

Use of microorganisms for the restoration of cultural heritage includes biocleaning and biomineralization.

Biocleaning: The process of utilizing microorganisms for restoration and conservation processes in order to eliminate organic materials is called biocleaning. Advantages of microbes over physicochemical methods lie in their utilization of substrate-specific enzymes which do not degrade complex substances and their ability to easily adapt to environmental conditions. The microbe selected for this purpose should be nonsporulating and non-pathogenic. Some examples from previous studies are removal of black scale, i.e. hydrated calcium sulphate and carbon residues in stone which are caused by hydrocarbons, sulphur dioxide and particulate matter. These were removed using bacteria *Desulfovibrio desulfuricans* (Cappitelli et al. 2006). Nitrates from marble were removed under anaerobic conditions using *Pseudomonas denitrificans*, *Paracoccus denitrificans* and *Pseudomonas stutzeri* (Cappitelli et al. 2009). *Pseudomonas stutzeri* was also employed to remove remains of organic matter from mural paints (Bosch-Roig et al. 2010).

Biomineralization: Modern ecological alternative for restoration is biomineralization also known as carbonatogenesis or calcite production which helps in restoration of cracks in statues or walls as selected bacteria are capable of mineralizing and filling these cracks upon feeding them with appropriate culture containing calcium salts in solution which produces microcrystals of calcium carbonate which allow restoration of damaged areas (Porqué 2018; Páramo Aguilera et al. 2011; Zammit et al. 2011; Rinaldi 2006; del Mar López-Miras et al. 2013; Calvo 1997). Calcite production occurs either autotrophically or heterotrophically by organisms like *Proteus*, *Myxococcus*, *Pantoea agglomerans*, *Pseudomonas* and *Bacillus*. The micro-environmental factors governing this process are pH, the concentration of calcium ions, inorganic carbon and presence of nucleation sites (Aguilera et al. 2015).

4.9 Cleaning Agents

Traditional natural cleaning agents include vinegar, wine, potash solutions, lemon juice, onion or garlic, sliced potatoes, physiological fluids like saliva or blood, bile fluids and urine (Wolbers 2000; Cremonesi 2004). The essential oils from medicinal plants among others are *Salvia officinalis*, *Artemisia absinthium*, *Mentha piperita*, *Lavandula angustifolia*, *Origanum compactum* and *Thymus vulgaris* (Walentowska and Foksowicz-Flaczyk 2013; Valdés-Pérez et al. 2016). These natural varnishes are used to enhance the visual properties of art through the saturation of colour. The same formulations which possess surface protection and hydrophobic properties are used in nearly every class of constituents and dynamic principles of modern cleaning agents.

During the start of the twentieth century, synthetic polymers were actively adopted thinking that they are extremely resilient to ageing and are easily removable. Synthetic polymers also undergo degradation similar to that of natural resins resulting in their solubility reduction in net solvents and significantly altering visual aspect (yellowing) of the art (Favaro et al. 2006). Owing to their adhesive properties, synthetic polymers have been utilized to re-adhere damaged or detached parts of the external layers of art. Some of the polymers extensively adopted for protection and consolidation of stone and wall paintings are epoxy polymers, silicone, acrylate and vinyl. Nevertheless application of synthetic adhesives brings about sturdy alteration in physicochemical properties like water capillarity, porosity, surface wettability and water vapour permeability of the original substrate (Horie 2013) which generate enhanced degradation in long term leading to loss of artefacts.

4.9.1 Cleaning Agents for Paper and Painting Heritage

Conservation practices for a paper usually include use of alkaline aqueous solutions for deacidification of paper (Magaudda 2004). However, application of aqueous solutions leads to swelling of cellulose fibres and leaching of associated compounds of paper like sizing, ink, etc. which demanded development of non-aqueous methods for deacidification. Methods adopted by restorers and conservators for combating paper degradation have been changing profoundly over the years (Sequeira et al. 2014). Seventy per cent ethanol was found to be very effective for short-term disinfection (Sequeira et al. 2017; Nittérus 2000). Further, ethylene oxide, quaternary ammonium salts, formaldehyde or alcohols were adopted to disinfect heritage objects over the last century (Rutala and Weber 1999). Nevertheless, common form of application by restorers through a typical spray does not remove all kinds of microorganisms giving a chance of recolonization frequently.

Analysis and understanding of fungi and their function attacking document-based heritage forms foundation for development of methods for cleaning and treatment (Guiamet et al. 2011; Pinzari et al. 2010; Michaelsen et al. 2010). From the available limited range of physical and chemical methods for disinfection of fungal damage, chemical treatments include fumigation with gases and using liquid biocides (Allsopp et al. 2004). Although topical chemical treatment is useful, efficacy is dependent on sensitivity of individual fungal species and tolerance of the surface to be treated. Gamma ray application can also result in cumulative depolymerization of cellulose and severe ageing characteristics (Adamo and Magaudda 2003).

Methods adopted for painting conservation include physical methods like UV, gamma rays and laser radiation, while mechanical methods adopted are removal with a tool or by hand. Further, chemical methods include antifungal agents like titanium dioxide and ZnO (Lugauskas and Jaskelevicius 2007). The most widely adopted method is adding antifungal agent in paint composition.

4.9.2 Modern Chemical Agents for the Conservation and Restoration of Cultural Heritage

A class of cleaning agents with combinations of natural agents is represented by surface-active agents or surfactants. Surfactants contain key components of saliva, bile fluids and other natural products. Ternary or binary systems composing surfactants, water and other additives belong to the realm of soft matter with characteristic length scales in nanometric domain that largely determines their interaction with complex micro-structured surfaces like pictorial layers. During the last decade and a half, tools of nanoscience, precisely from colloid and surface science were studied to overcome limitations of traditional restoration methods through providing new cultural framework and innovative methodologies. As a result, these numerous consistent, inexpensive and easy-to-use tools have been conceived and presented to conservators consisting of novel broad treatments exhibiting high compatibility on application to original artistic substrates and long-term durability (Nicolai et al. 2009).

In order to predict, asses or improve the efficacy of a cleaning agent, its mechanism of interaction with the material to be removed on heritage is to be understood to the precise nanoscale. This makes it clear that synergy between nanoscience, conservation science and surface science is key for the development of powerful yet environmentally friendly and mild cleaning tools. An emerging field currently in development for conservation of cultural heritage is nanotechnology. European Commission has proposed the project FP7 NANOFORART whose main objective is to conserve and preserve both movable and immovable heritages. Applications of this project have reported utilization of nanoparticles of zinc oxide for fungal biofilm control and application of nanosilver-coated cotton fibres for antimicrobial textile finishing (Gambino et al. 2017; Di Salvo 2014; Baglioni et al. 2012).

Nanoparticles in terms of nanosols, alcohol dispersions and powders have been applied for prevention of degradation of paper, painting, wood and stone heritage during the last decade (Poggi et al. 2016). Effectiveness of nanoparticles lies in their capability to interact with proteins and DNA and pass through the cell membranes of microbes. Silver nanoparticles are one of the most widely used offering high antimicrobial properties even at lower concentrations and multiple mechanisms of action (Rai et al. 2009; Kim et al. 2007; Sondi and Salopek-Sondi 2004; Feng et al. 2000). Toxicity of silver to cells of higher animals which can disturb ecological balance is of concern for its use (Blaser et al. 2008; Hsin et al. 2008). Other alternative nanoparticles include nanostructured zinc oxide, which is an antifungal compound having biocompatibility and is safe, and nanometric form of TiO_2 characterized by low cytotoxicity and broad-spectrum biocidal activity, which has been found effective against fungi (Ruffolo et al. 2010; Janaki et al. 2015; Munafo et al. 2015).

Tuneable physicochemical properties and flexible nature of nanostructured inorganic materials allow us to obtain an archive of materials in various shapes, sizes and surface properties (Tang et al. 2012; He et al. 2016). With reference to conservation of documentary heritage, high surface area of nanoparticles effectively covers and penetrates paper into network of cellulose fibres neutralizing acids and thus protecting paper from cellulose hydrolysis. Removal of hydrophobic material from hydrophilic fresco painting was attained using microemulsions (Ferroni et al. 1992; De Gennes and Taupin 1982). Better results were obtained when aqueous solution of sodium dodecyl sulphate/1-PeOH micelles is added with 8% of ethyl acetate and 8% propylene carbonate, which allowed complete cleaning of the surface in a few minutes (Carretti et al. 2003).

Hydroxide nanoparticles are used for consolidation of wall paintings and are also used to control pH of documentary heritage and wood avoiding catalytic cleavage of –glycosidic bond occurring in presence of transition metal ions, which is a new perspective towards conservation of an enormous documentary heritage (Baglioni et al. 2012).

Among systems that are explicitly tailored for the conservation of heritage, nanoparticulate inorganic sols called nanosols, colloidal silica and alkoxysilane paly key role in conservation of stone and wood (Mahltig et al. 2008; Stepien et al. 1970; Wheeler et al. 2003). Further, owing to their high surface-to-volume ratio, nanosols are metastable and usually hydrolysed to form 3D xerogel networks which enhance mechanical properties and resistance towards fire, water and insect or microbial attack of stone or wood objects.

In comparison with traditional pure solvents for cleaning, advantages of microemulsions rely on enhanced soil removal and hydrophobic material confinement inside the oil microemulsion droplets (Borgioli et al. 1995). Further aqueous amphiphilic systems result in a significant reduction in amount of organic solvents for typical cleaning procedure, which reduces system's toxicity and environmental impact. Surface cleaning by microemulsion is through interaction with the polymer where fraction of solvent migrates from microemulsion droplets to the coating, and as the organic fraction is lost, the size of micelle decreases leading to detachment of swollen polymer from surface resulting in cleaning (Baglioni et al. 2012).

The systems of solvent gels enable control of components like cosolvents, detergents and enzymes by limiting the evaporation of solvents apart from their penetration into the artwork to be cleaned. Owing to the problems associated with requirement of solvents for removal of residues of solvent gels (Stulik et al. 2004), gels with higher solvent retention capability are required for cleaning hydrophilic artworks. A system with high control over water confinement was obtained with a semi-interpenetrating poly(N-vinyl-1-pyrrolidone) (PVP) and poly(2-hydroxyethyl methacrylate) (pHEMA) structure. Another class is high-viscosity polymeric dispersions (HVPDs), which do not truly gel with reference to their rheological behaviour. The key feature of these systems is the possibility of their removal by peeling without use of neat solvents. These types of highly retentive supporting gels coupled with nanostructured amphiphilic systems offer a novel palette of cleaning tools for conservators. These systems come with advantages of requiring selective, easy and controlled removal of undesired layers of grime, dirt and polymers.

Further, nanoparticles of alkaline earth metal hydroxides are considered to be the most durable and reliable systems. Calcium and magnesium hydroxides have proved to be excellent compounds for deacidification of cellulosic heritage (Salvadori and Dei 2001; Ambrosi et al. 2001). Owing to the drawbacks of application of alkaline aqueous solutions to water-sensitive heritage and probability of favouring alkaline depolymerization of cellulose, non-aqueous methods, dispersions of oxides and carbonate precursors are proposed which overcome these drawbacks (Baty et al. 2010). Alkaline earth metal hydroxide nanoparticles dispersed in alcohols have recently proved to be competent deacidification system for not only paper but also promising tool for archaeological wood (Chelazzi et al. 2006).

A combination of barium and calcium hydroxide nanoparticle dispersion has proved to be extremely competent in the consolidation of mural paintings that were profoundly contaminated by salts of chlorides and sulphates in the Mesoamerican area (Chelazzi et al. 2013). Moreover, these are very low in toxicity over barium salts.

4.10 Conclusion and Future Prospects

Apart from the natural ageing process leading to deterioration, environmental factors are understood to play a detrimental effect on the cultural heritage, which provide favourable conditions for microbial colonization. Studying deterioration of cultural heritage by fungi is significant not only for understanding effects of drastically changing environmental conditions but also to comprehend the phenomenon of physical and chemical deterioration. Knowledge on chemical composition of cultural heritage from paintings to stone provides an understanding of type of microbial colonies that might be favoured leading to their degradation. Biofilm formation is a microbial strategy to adapt to adverse conditions for their growth. Hence, an understanding of the microbial species forming biofilms and their characteristics is mandatory for controlling threat to objects of interest to allow next generation in knowing this legacy.

During the last decade and a half, nanoscience has contributed to conservation of immovable and movable cultural heritage. Using nanomaterials that are compatible with heritage under treatment results in stability for long term that has economic and societal benefits. From the literature, it can be assumed that research and development in the science of cultural heritage are far from being mature. Chief future possible scenarios comprise of "green" surfactant-based, self-assembled systems, organogels, nanoparticle dispersions in apolar solvents towards deacidification of water-sensitive substrates, waterless cleaning microemulsions and hybrid organic-inorganic nanocomposites. Additionally efforts for development of novel formulation with use of biodegradable non-ionic surfactants with the aim of producing environmentally friendly yet effective systems are needed.

References

Abdel-Kareem O (2010) Monitoring, controlling and prevention of the fungal deterioration of textile artifacts in the museum of Jordanian heritage. Mediter Archaeol Archaeom 10:85–96

Adamo M, Magaudda G (2003) Susceptibility of printed paper to attack of chewing insects after gamma irradiation and ageing. Restaurator 24:95–105

Agarwal PN, Puvathingal JM (1969) Microbiological deterioration of woolen materials. Textile Res J 39:38–42

Agrawal OP, Dhawan S, Garg KL, Shaheen F, Pathak N, Misra A (1988) Study of biodeterioration of the Ajanta wall paintings. Int Biodeterior 24:121–129

Aguilera LAP, Zapata JAN, Morales BOO (2015) La bioprecipitación de carbonato de calcio por la biota nativa como un método de restauración. Nexo Revista Científica 28:25–40

Allsopp D, Seal KJ, Gaylarde CC (2004) Introduction to biodeterioration. Cambridge University Press

Ambrosi M, Dei L, Giorgi R, Neto C, Baglioni P (2001) Colloidal particles of Ca (OH) 2: properties and applications to restoration of frescoes. Langmuir 17:4251–4255

Arai H (1984) Microbiological studies on the conservation of mural paintings in tumuli. In International symposium on the conservation and restoration of cultural property. Conservation and restoration of mural paintings (I). November 17–21, 1983, Tokyo, Japan, pp 117–124

Arai H (2000) Foxing caused by fungi: twenty-five years of study. Int Biodeterior Biodegradation 46:181–188

Area MC, Ceradame H (2011) Paper aging and degradation: recent findings and research methods. BioResources 6(4):5307–5337

Baglioni M, Giorgi R, Berti D, Baglioni P (2012) Smart cleaning of cultural heritage: a new challenge for soft nanoscience. Nanoscale 4:42–53

Bastian F, Alabouvette C (2009) Lights and shadows on the conservation of a rock art cave: the case of Lascaux Cave. Int J Speleol 38(6):55

Baty JW, Maitland CL, Minter W, Hubbe MA, Jordan-Mowery SK (2010) Deacidification for the conservation and preservation of paper-based works: a review. Bioresources 5:1955–2023

Bech-Andersen J, Elborne SA, Goldie F, Singh J, Singh S, Walker B (1993). The true dry rot fungus (*Serpula lacrymans*) found in the wild in the forests of the Himalayas. Document-the International Research Group on Wood Preservation (Sweden)

Berner M, Wanner G, Lubitz W (1997) A comparative study of the fungal flora present in medieval wall paintings in the chapel of the castle Herberstein and in the parish church of St Georgen in Styria, Austria. Int Biodeterior Biodegradation 40:53–61

Bertolin C (2019) Preservation of cultural heritage and resources threatened by climate change. Geoscience 9(6)

Blaser SA, Scheringer M, MacLeod M, Hungerbühler K (2008) Estimation of cumulative aquatic exposure and risk due to silver: contribution of nano-functionalized plastics and textiles. Sci Total Environ 390:396–409

Błyskal B (2005) The Influence of Dyes on the Degree of Biodeterioration of a Woollen Textile. Doctoral dissertation, PhD Dissertation. Krakow University of Economics, Faculty of Commodity Science, Krakow, Poland (in Polish)

Błyskal B (2009) Fungi utilizing keratinous substrates. Int Biodeterior Biodegradation 63:631–653

Błyskal B (2014) Gymnoascus arxii's potential in deteriorating woollen textiles dyed with natural and synthetic dyes. Int Biodeterior Biodegradation 86:349–357

Błyskal B (2015) Fungal deterioration of a woollen textile dyed with cochineal. J Cult Herit 16:32–39

Bock E (1993) The microbiology of masonry biodeterioration. J Appl Bacteriol 74:503–514

Borgioli L, Caminati G, Gabrielli G and Ferroni E (1995) Removal of hydrophobic impurities from pictorial surfaces by means of heterogeneous systems. Sci Technol Cult Herit J "Comitato Nazionale per la Scienza e la Tecnologia dei Beni Culturali", CNR, 4:67–74

Borrego-Alonso S (2015) Los biocidas vegetales en el control del biodeterior o del patrimonio documental. Perspectivas e impacto. Revista CENIC. Cienc Biol 46:259–269

Bosch-Roig P, Regidor-Ros JL, Soriano-Sancho P, Domenech-Carbo MT, Montes-Estelles RM (2010) Ensayos de biolimpieza con bacterias en pinturas murales. Arche 4-5:115–124

Calvo A (1997) Conservación y restauración: materiales, técnicas y procedimientos: de la A a la Z

Cameotra SS, Dakal TC (2012) Carbonatogenesis: microbial contribution to the conservation of monuments and artwork of stone. Conserv Sci Cult Herit 12:79–108

Cappitelli F, Sorlini C (2005) From papyrus to compact disc: The microbial deterioration of documentary heritage. Crit Rev Microbiol 31(1):1–10

Cappitelli F, Zanardini E, Ranalli G, Mello E, Daffonchio D, Sorlini C (2006) Improved methodology for bioremoval of black crusts on historical stone artworks by use of sulfate-reducing bacteria. Appl Environ Microbiol 72:3733–3737

Cappitelli F, Fermo P, Vecchi R, Piazzalunga A, Valli G, Zanardini E, Sorlini C (2009) Chemical–physical and microbiological measurements for indoor air quality assessment at the Ca'Granda Historical Archive, Milan (Italy). Water Air Soil Pollut 201:109–120

Carretti E, Dei L, Baglioni P (2003) Solubilization of acrylic and vinyl polymers in nanocontainer solutions. Application of microemulsions and micelles to cultural heritage conservation. Langmuir 19:7867–7872

Castrillón L, Marañón E, Fernández-Nava Y, Ormaechea P, Quiroga G (2013) Thermophilic co-digestion of cattle manure and food waste supplemented with crude glycerin in induced bed reactor (IBR). Bioresour Technol 136:73–77

Castro RDT, López ALB (2013) Enfoque químico del deterioro y biodeterioro de rocas calcáreas conformantes de monumentos patrimoniales de importancia histórica y cultural. Revista Luna Azul:247–284

Chahinian H, Nini L, Boitard E, Dubès JP, Comeau LC, Sarda L (2002) Distinction between esterases and lipases: a kinetic study with vinyl esters and TAG. Lipids 37:653–662

Chelazzi, D., Giorgi, R., and Baglioni, P. (2006). Nanotechnology for Vasa Wood De-Acidification. In Macromolecular symposia (Vol. 238, 1, pp. 30–36). Weinheim: WILEY-VCH Verlag

Chelazzi D, Poggi G, Jaidar Y, Toccafondi N, Giorgi R, Baglioni P (2013) Hydroxide nanoparticles for cultural heritage: consolidation and protection of wall paintings and carbonate materials. J Colloid Interface Sci 92:42–49

Cheng HN, Gross RA (2003) Biocatalysis in polymer science: an overview. ACS Symp Ser 840:1–33

Ciferri O (1999) Microbial degradation of paintings. Appl Environ Microbiol 65:879–885

Ciferri O, Nola P, Seves AM, Seves A, Meloni S, Oddone M, Gallone A (1996) L'affresco di Sant'Agata al Monte di Pavia: ricerche ed analisi per il restauro. Memorie dell'Istituto Lombardo. Accademia di Scienze e Lettere 40:167–251

Ciferri O, Tiano P, Mastromei G (eds) (2000) Of microbes and art: the role of microbial communities in the degradation and protection of cultural heritage. Springer, Boston

Claydon N (1984) Fusarium as an insect pathogen. In: The applied mycology of Fusarium. Cambridge University Press, Cambridge, UK, pp 117–128

Corte AM, Ferroni A, Salvo VS (2003) Isolation of fungal species from test samples and maps damaged by foxing, and correlation between these species and the environment. Int Biodeterior Biodegradation 51:167–173

Costerton JW, Cheng KJ, Geesey GG, Ladd TI, Nickel JC, Dasgupta M, Marrie TJ (1987) Bacterial biofilms in nature and disease. Annu Rev Microbiol 41:435–464

Coutinho ML, Miller AZ, Phillip A, Mirão J, Dias L, Rogerio-Candelera MA, Macedo MF (2019) Biodeterioration of majolica glazed tiles by the fungus *Devriesia imbrexigena*. Constr Build Mater 212:49–56

Cremonesi P (2004) L'uso dei tensioattivi e chelanti nella pulitura di opere policrome

Cuetos MPG (2012) El patrimonio cultural. Conceptos básicos (Vol. 207). Universidad de Zaragoza

Dakal TC, Arora PK (2012) Evaluation of potential of molecular and physical techniques in studying biodeterioration. Rev Environ Sci Biotechnol 11(1):71–104

Dakal TC, Cameotra SS (2012) Microbially induced deterioration of architectural heritages: routes and mechanisms involved. Environ Sci Eur 24:36

De Gennes PG, Taupin C (1982) Microemulsions and the flexibility of oil/water interfaces. J Phys Chem 86:2294–2304

de Vries RP, Visser JAAP (2001) *Aspergillus* enzymes involved in degradation of plant cell wall polysaccharides. Microbiol Mol Biol Rev 65:497–522

del Mar López-Miras M, Martín-Sánchez I, Yebra-Rodríguez Á, Romero-Noguera J, Bolívar-Galiano F, Ettenauer J, Piñar G (2013) Contribution of the microbial communities detected on an oil painting on canvas to its biodeterioration. PLoS One 8:e80198

Di Bonaventura MP, Del Gallo M, Cacchio P, Ercole C, Lepidi A (1999) Microbial formation of oxalate films on monument surfaces: bioprotection or biodeterioration? Geomicrobiol J 16:55–64

Di Salvo S (2014) Nanotechnology for cultural heritage. Sci Technol Soc Int J 2:28–32

Domenech-Carbo MT (2008) Novel analytical methods for characterising binding media and protective coatings in artworks. Anal Chim Acta 621:109–139

Domingues JA, Bonelli N, Giorgi R, Fratini E, Gorel F, Baglioni P (2013) Innovative hydrogels based on semi-interpenetrating p(HEMA)/PVP networks for the cleaning of water-sensitive cultural heritage artifacts. Langmuir 29:2746–2755

Dupont J, Jacquet C, Dennetiere B, Lacoste S, Bousta F, Orial G, Roquebert MF (2007) Invasion of the French Paleolithic painted cave of Lascaux by members of the *Fusarium solani* species complex. Mycologia 99:526–533

Estela CRL, Alejandro PR (2012) Biofilms: a survival and resistance mechanism of microorganisms. In Antibiotic Resistant Bacteria-A Continuous Challenge in the New Millennium. IntechOpen

Ettenauer J, Sterflinger K, Piñar G (2010) Cultivation and molecular monitoring of halophilic microorganisms inhabiting an extreme environment presented by a salt-attacked monument. Int J Astrobiol 9:59–72

Ettenauer JD, Piñar G, Lopandic K, Spangl B, Ellersdorfer G, Voitl C, Sterflinger K (2012) Microbes on building materials—evaluation of DNA extraction protocols as common basis for molecular analysis. Sci Total Environ 439:44–53

Fabbri AA, Ricelli A, Brasini S, Fanelli C (1997) Effect of different antifungals on the control of paper biodeterioration caused by fungi. Int Biodeterior Biodegradation 39:61–65

Favaro M, Mendichi R, Ossola F, Russo U, Simon S, Tomasin P, Vigato PA (2006) Evaluation of polymers for conservation treatments of outdoor exposed stone monuments. Part I: photo-oxidative weathering. Polym Degrad Stabil 9:3083–3096

Feng QL, Wu J, Chen GQ, Cui FZ, Kim TN, Kim JO (2000) A mechanistic study of the antibacterial effect of silver ions on *Escherichia coli* and *Staphylococcus aureus*. J Biomed Mater Res 52:662–668

Ferroni E, Baglioni P (1984) Experiments on a proposed method for restoration of sulphated frescos. In Scientific methodologies applied to works of art. Proceedings of the symposium, Florence, Italy 2–5 May 1984. pp 108–109

Ferroni E, Malaguzzi Valerj V, Rovida G (1969) Experimental study by diffraction of heterogeneous systems as a preliminary to the proposal of a technique for the restoration of gypsum polluted murals

Ferroni E, Gabrielli G, Caminati G (1992) ASportazione di materiali idrofobi da superfici pittoriche murali mediante solubilizzazione in sistemi dispersi. La Cappella Brancacci, la scienza per Masaccio, Masolino e Filippino Lippi, ed. Olivetti, Quaderni del restauro

Flemming HC (1998) Relevance of biofilms for the biodeterioration of surfaces of polymeric materials. Polym Degrad Stabil 59:309–315

Florian MLE (2002) Fungal facts. Archetype, London

Gallo F (1965) Nota sulla conservazione dei microfilm. Ist Patologia del Libro Boll (1–4)

Gambino M, Ahmed MAAA, Villa F, Cappitelli F (2017) Zinc oxide nanoparticles hinder fungal biofilm development in an ancient Egyptian tomb. Int Biodeterior Biodegrad 122:92–99

Gargani G (1968) Fungus contamination of Florence art masterpieces before and after the 1966 disaster. In: Biodeterioration of materials, microbiological and allied aspects. Elsevier, Amsterdam, pp 252–257

Gaylarde CC, Morton LG (1999) Deteriogenic biofilms on buildings and their control: a review. Biofouling 14(1):59–74

Gettens RJ, Pease M, Stout GL (1941) The problem of mold growth in paintings. In: Technical studies in the field of the fine arts (Vol. 9). p 127

Geweely NS (2006) Non-toxic fumigation and alternative control techniques against fungal colonization for preserving archaeological oil painting. Int J Bot 2:353–362

Giacobini C, Firpi M (1981) Problemi di microbiologia nei dipinti su tela. In Atti del convegno sul restauro delle opere d'arte. Firenze, 2–7 novembre 1976, pp 203–212

Gorbushina AA, Petersen K (2000) Distribution of microorganisms on ancient wall paintings as related to associated faunal elements. Int Biodeterior Biodegrad 46(4):277–284

Gorbushina AA, Heyrman J, Dornieden T, Gonzalez-Delvalle M, Krumbein WE, Laiz L, Petersen K, Saiz-Jimenez C, Swings J (2004) Bacterial and fungal diversity and biodeterioration problems in mural painting environments of St. Martins church (Greene–Kreiensen, Germany). Int Biodeterior Biodegrad 53(1):13–24

Greif MD, Currah RS (2007) Patterns in the occurrence of saprophytic fungi carried by arthropods caught in traps baited with rotted wood and dung. Mycologia 99(1):7–19

Guglielminetti M, Morghen CDG, Radaelli A, Bistoni F, Carruba G, Spera G, Caretta G (1994) Mycological and ultrastructural studies to evaluate biodeterioration of mural paintings. Detection of fungi and mites in frescos of the monastery of St damian in assisi. Int Biodeterior Biodegrad 33(3):269–283

Guiamet, P., Borrego, S., Lavin, P., Perdomo, I., and de Saravia, S. G. (2011) Biofouling and biodeterioration in materials stored at the historical archive of the Museum of La Plata, argentine and at the National Archive of the Republic of Cuba. Colloids Surf B: Biointerfaces, 85(2), 229–234

Gunde-Cimerman N, Zalar P, Jeram S (1998) Mycoflora of cave cricket *Troglophilus neglectus* cadavers. Mycopathologia 141(2):111–114

Gutarowska B, Pietrzak K, Machnowski W, Milczarek JM (2017) Historical textiles–a review of microbial deterioration analysis and disinfection methods. Text Res J 87(19):2388–2406

He Y, Ingudam S, Reed S, Gehring A, Strobaugh TP, Irwin P (2016) Study on the mechanism of antibacterial action of magnesium oxide nanoparticles against foodborne pathogens. J Nanobiotechnol 14(1):54

Hoang CP, Kinney KA, Corsi RL, Szaniszlo PJ (2010) Resistance of green building materials to fungal growth. Int Biodeterior Biodegrad 64(2):104–113

Horie CV (2013) Materials for conservation. Routledge

Hsin YH, Chen CF, Huang S, Shih TS, Lai PS, Chueh PJ (2008) The apoptotic effect of nanosilver is mediated by a ROS-and JNK-dependent mechanism involving the mitochondrial pathway in NIH3T3 cells. Toxicol Lett 179(3):130–139

Hueck HJ (1965) The biodeterioration of materials as a part of hylobiology. Mater Org 1(1):5–34

Humber RA, Hansen KS (2005) ARSEF Index: Host-by Fungus. Fungus, Host order, Host family, Host genus, Host species Arsef (28 September 2005). USDA-ARS Plant Protection Research Unit US Plant, Soil and Nutrition Lab. NY (USA) http://arsef.fpsnl.cornell.edu/mycology/catalogs/Indices/hostfung.pdf

ICOMOS International Committee on Cultural Tourism (2002) ICOMOS international cultural tourism charter: principles and guidelines for managing tourism at places of cultural and heritage significance. International Council on Monuments and Sites, ICOMOS International Cultural Tourism Committee

Inoue M, Koyano M (1991) Fungal contamination of oil paintings in Japan. Int Biodeterior 28 (1–4):23–35

Ionita I (1973) Contributions to the study of the biodeterioration of the works of art and historical monuments. IV. Fungi involved in the deterioration of mural painting from the monasteries of Moldavia. Revue roumaine de biologie. Serie de botanique

James TY, Kauff F, Schoch CL, Matheny PB, Hofstetter V, Cox CJ et al (2006) Reconstructing the early evolution of Fungi using a six-gene phylogeny. Nature 443(7113):818

Janaki AC, Sailatha E, Gunasekaran S (2015) Synthesis, characteristics and antimicrobial activity of ZnO nanoparticles. Spectrochim Acta A Mol Biomol Spectrosc 144:17–22

Jeffries P (1986) Growth of *Beauvaria alba* on mural paintings in Canterbury cathedral. Int Biodeterior 22(1):11–13

Jerusik RJ (2010) Fungi and paper manufacture. Fungal Biol Rev 24(1–2):68–72

Joshi VK, Attri D (2005) Solid state fermentation of apple pomace for the production of value added products. Pollut Urban Ind Environ 180

Jurado V, Sanchez-Moral S, Saiz-Jimenez C (2008) Entomogenous fungi and the conservation of the cultural heritage: A review. Int Biodeterior Biodegradation 62(4):325–330

Kaarakainen P, Rintala H, Vepsäläinen A, Hyvärinen A, Nevalainen A, Meklin T (2009) Microbial content of house dust samples determined with qPCR. Sci Total Environ 407(16):4673–4680

Karamanos Y (1997) Endo-N-acetyl-β-D-glucosaminidases and their potential substrates: structure/function relationships. Res Microbiol 148(8):661–671

Kim JS, Kuk E, Yu KN, Kim JH, Park SJ, Lee HJ et al (2007) Antimicrobial effects of silver nanoparticles. Nanomedicine 3(1):95–101

Koestler RJ, Koestler VH (eds) (2003) Art, biology, and conservation: biodeterioration of works of art. Metropolitan museum of art, New York

Kowalik R (1980) Chapter 4: Microbiodeterioration of library materials. Part 2. Microbiodecomposition of basic organic library materials. Restaurator 4(3–4):135–220

Kubátová A, Dvorák L (2005) Entomopathogenic fungi associated with insect hibernating in underground shelters. Czech Mycol 57(3/4):221

Kunert J (1989) Biochemical mechanism of keratin degradation by the actinomycete *Streptomyces fradiae* and the fungus *Microsporum gypseum*: a comparison. J Basic Microbiol 29(9):597–604

Leatherdale D (1970) The arthropod hosts of entomogenous fungi in Britain. Entomophaga 15 (4):419–435

López-Miras M, Piñar G, Romero-Noguera J, Bolivar-Galiano FC, Ettenauer J, Sterflinger K, Martin-Sanchez I (2013) Microbial communities adhering to the obverse and reverse sides of an oil painting on canvas: Identification and evaluation of their biodegradative potential. Aerobiologia 29(2):301–314

Lugauskas A, Jaskelevicius B (2007) Microbiological destruction of constructional and decoration materials of buildings. Materi Sci (Medžiagotyra) 13(1):70–73

Magaudda G (2004) The recovery of biodeteriorated books and archive documents through gamma radiation: some considerations on the results achieved. J Cult Herit 5(1):113–118

Mahltig B, Swaboda C, Roessler A, Böttcher H (2008) Functionalising wood by nanosol application. J Mater Chem 18(27):3180–3192

Maina UM, Galadima IB, Gambo FM, Zakaria D (2018) A review on the use of entomopathogenic fungi in the management of insect pests of field crops. J Entomol Zool Stud 6(1):27–32

Mason-Williams A (1965) The growth of fungi in caves in Great Britain. Stud Speleol 1 (2–3):96–99

Matočec N, Ozimec R (2001) Observations on Cordyceps riverae (Hypocreales, Ascomycota) in Croatian caves. Nat Croat Periodicum Musei Hist Natur Croat 10(3):197–206

Meier C, Petersen K (2006) Schimmelpilze auf Papier, ein Handbuch für Restauratoren: biologische Grundlagen, Erkennung, Behandlung und Prävention

Melo D, Sequeira SO, Lopes JA, Macedo MF (2019) Stains versus colourants produced by fungi colonising paper cultural heritage: a review. J Cult Herit 35:161–182

Menei E (1990) Le papyrus

Mesquita N, Portugal A, Videira S, Rodríguez-Echeverría S, Bandeira AML, Santos MJA, Freitas H (2009) Fungal diversity in ancient documents. A case study on the archive of the University of Coimbra. Int Biodeterior Biodegrad 63(5):626–629

Michaelsen A, Piñar G, Pinzari F (2010) Molecular and microscopical investigation of the microflora inhabiting a deteriorated Italian manuscript dated from the thirteenth century. Microb Ecol 60(1):69–80

Milicia MT (2014) Lombroso e il brigante: storia di un cranio conteso. Salerno

Mondal S, Baksi S, Koris A, Vatai G (2016) Journey of enzymes in entomopathogenic fungi. Pac Sci Rev A Nat Sci Eng 18(2):85–99

Monte M (2003) Oxalate film formation on marble specimens caused by fungus. J Cult Herit 4 (3):255–258

Morton LHG, Surman SB (1994) Biofilms in biodeterioration—a review. Int Biodeterior Biodegrad 34(3–4):203–221

Munafo P, Goffredo GB, Quagliarini E (2015) TiO2-based nanocoatings for preserving architectural stone surfaces: an overview. Constr Build Mater 84:201–218

Neilands JB (1995) Siderophores: structure and function of microbial iron transport compounds. J Biol Chem 270(45):26723–26726

Nicolai P, Lombardi D, Tomenzoli D, Villaret AB, Piccioni M, Mensi M, Maroldi R (2009) Fungus ball of the paranasal sinuses: Experience in 160 patients treated with endoscopic storia di un cranio conteso surgery. Laryngoscope 119(11):2275–2279

Nielsen KF, Holm G, Uttrup LP, Nielsen PA (2004) Mould growth on building materials under low water activities. Influence of humidity and temperature on fungal growth and secondary metabolism. Int Biodeterior Biodegrad 54(4):325–336

Nittérus M (2000) Fungi in archives and libraries. Restaurator 21(1):25–40

Nugari MP, Priori GF (1985, September) Resistance of acrylic polymers to microorganisms-first part. 5th international congress on deterioration and conservation of stone. Lausanne, pp 25–27

Nugari MP, Realini M, Roccardi A (1993) Contamination of mural paintings by indoor airborne fungal spores. Aerobiologia 9(2):131–139

Nyuksha YP (1983) Some special cases of biological deterioration of books. Restaurator 5 (3–4):177–182

Obidi OF, Aboaba OO, Makanjuola MS, Nwachukwu SCU (2009) Microbial evaluation and deterioration of paints and paint-products. J Environ Biol 30(5):835

Olmedo-Juárez A, Rojo-Rubio R, Zamilpa A, De Gives PM, Arece-García J, López-Arellano ME, von Son-de Fernex E (2017) In vitro larvicidal effect of a hydroalcoholic extract from *Acacia cochliacantha* leaf against ruminant parasitic nematodes. Vet Res Commun 41(3):227–232

Pangallo D, Chovanova K, Šimonovičová A, Ferianc P (2009) Investigation of microbial community isolated from indoor artworks and air environment: identification, biodegradative abilities, and DNA typing. Can J Microbiol 55(3):277–287

Pankhurst ES, Davies MJ, Blake HM (1972) The ability of polymers or materials containing polymers to provide a source of carbon for selected microorganisms. In Biodeterioration of materials. Proceedings of the 2nd international biodeterioration symposium. Lunteren, The Netherlands, 13th–18th September 1971 (pp. 76–90)

Páramo Aguilera LA, Zapata N, De la Cruz E (2011) Aislamiento e identificación de microorganismos en biopelículas provenientes del Castillo de Chapultepec, Ciudad de México. Nexo Revista Cient 24(2):1–9

Pekhtasheva, E., Neverov, A., Kubica, S., Zaikov, G. (2012). Biodegradation and biodeterioration of some natural polymers

Péreza JR, Martínezb PG Lights and shadows over the implementation of the Recommendation on the Historic Urban Landscape: "Managing change" in Ballarat and Cuenca using a radical approach focused on values and authenticity

Petersen K, Grote G, Krumbein WE (1988) Biotransfer of metals by fungi isolated from rock. In VIth International Congress on deterioration and conservation of stone. Supplement.= VIe Congrès International sur l'altération et la conservation de la pierre. Torun, 12-14.09. 1988, pp 111–119

Petrov P, Petrova E, Tchorbanov B, Tsvetanov CB (2007) Synthesis of biodegradable hydroxyethylcellulose cryogels by UV irradiation. Polymer 48(17):4943–4949

Petushkova JP, Lyalikova NN (1986) Microbiological degradation of lead-containing pigments in mural paintings. Stud Conserv 31(2):65–69

Pinzari F, Montanari M (2011) Mould growth on library materials stored in compactus-type shelving units. In: Sick building syndrome. Springer, Berlin, Heidelberg, pp 193–206

Pinzari F, Cialei V, Barbabietola N (2010) Measurement of the fungal deteriorating potential in the dust of indoor environments. e-Preservation Sci J 7:29–34

Poggi G, Toccafondi N, Chelazzi D, Canton P, Giorgi R, Baglioni P (2016) Calcium hydroxide nanoparticles from solvothermal reaction for the deacidification of degraded waterlogged wood. J Colloid Interface Sci 473:1–8

Porqué es importante la conservación del patrimonio cultural? Available from: https://sabiendomsdenuestropatrimonio cultural.wordpress.com. Accessed: 7 May 2018

Poyatos Jiménez F (2007) Procesos de biodeterioro en pinturas sobre lienzo del Museo de Bellas Artes de Granada: examen visual y gráfico. Universidad de Granada, Granada

Prompayuk S, Chairattananon P (2016) Preservation of cultural heritage community: cases of Thailand and developed countries. Procedia Soc Behav Sci 234:239–243

Rai M, Yadav A, Gade A (2009) Silver nanoparticles as a new generation of antimicrobials. Biotechnol Adv 27(1):76–83

Ravikumar HR, Rao SS, Karigar CS (2012) Biodegradation of paints: a current status. Indian J Sci Technol 5(1):1977–1987

Rebricova NL (1991) Some ecological aspects of protection of old Russian wall paintings from microbiological deterioration. In International conference on biodeterioration of cultural property. 20–25 February, 1989. Preprints, Vol. I, pp 25–37

Rinaldi A (2006) Saving a fragile legacy. EMBO Rep 7(11):1075–1079

Rivera CM (2018) Patrimonio Cultural Tangible e Intangible. Available from: https://manuelrivera23.wordpress.com/blog/revista-ajayu/patrimonio-culturaltangible-e-intangible/

Rivera LEC, Ramos AP, Desgarennes MDCP (2010) Importancia de las biopelículas en la práctica médica. práctica médica 54(1):14–24

Roman C, Diaconescu R, Scripcariu L, Grigoriu A (2013) Biocides used in preservation, restoration and conservation of the paper. Eur J Sci Theol 9(4):263–271

Ruffolo SA, La Russa MF, Malagodi M, Rossi CO, Palermo AM, Crisci GM (2010) ZnO and $ZnTiO_3$ nanopowders for antimicrobial stone coating. Appl Phys A 100(3):829–834

Ruga L, Bonofiglio T, Orlandi F, Romano B, Fornaciari M (2008) Analysis of the potential fungal biodeteriogen effects in the "Doctorate Library" of the University of Perugia, Italy. Grana 47 (1):60–69

Ruga L, Orlandi F, Romano B, Fornaciari M (2015) The assessment of fungal bioaerosols in the crypt of St. Peter in Perugia (Italy). Int Biodeterior Biodegrad 98:121–130

Rutala WA, Weber DJ (1999) Infection control: the role of disinfection and sterilization. J Hosp Infect 43:S43–S55

Saarela M, Alakomi HL, Suihko ML, Maunuksela L, Raaska L, Mattila-Sandholm T (2004) Heterotrophic microorganisms in air and biofilm samples from Roman catacombs, with special emphasis on actinobacteria and fungi. Int Biodeterior Biodegrad 54(1):27–37

Saiz-Jimenez C (1993) Deposition of airborne organic pollutants on historic buildings. Atmos Environ Part B Urban Atmos 27(1):77–85

Saiz-Jimenez C, Samson RA (1981) Biodegradación de obras de arte. Hongos implicados en la degradación de los frescos del monasterio de la Rábida (Huelva). Bot Macaronesica 8:255–264

Salvadori B, Dei L (2001) Synthesis of Ca (OH) 2 nanoparticles from diols. Langmuir 17 (8):2371–2374

Salvadori O, Municchia AC (2016) The role of fungi and lichens in the biodeterioration of stone monuments. Open Conf Proc J 7(1):39

Savulescu A, Ionita I (1971) Contributions to the study of the biodeterioration of the works of art and historic monuments, i. Species of fungi isolated from frescoes. Rev Roum Biol 16 (3):201–206

Schmitt JA (1974) The microecology of mold growth. J Paint Technol 46(599):59–64

Sequeira SO, Cabrita EJ, Macedo MF (2014) Fungal biodeterioration of paper: how are paper and book conservators dealing with it? An international survey. Restaurator Int J Preservation Library Arch Mater 35(2):181–199

Sequeira SO, Phillips AJ, Cabrita EJ, Macedo MF (2017) Ethanol as an antifungal treatment for paper: short-term and long-term effects. Stud Conserv 62(1):33–42

Seves A, Romanò M, Maifreni T, Sora S, Ciferri O (1998) The microbial degradation of silk: a laboratory investigation. Int Biodeterior Biodegrad 42(4):203–211

Shahid AA, Rao AQ, Bakhsh A, Husnain T (2012) Entomopathogenic fungi as biological controllers: new insights into their virulence and pathogenicity. Arch Biol Sci 64(1):21–42

Sileo M, Gizzi FT, Masini N (2015) Monitoraggio microclimatico: passato, presente e prospettive future. Salvaguardia, Conservazione e Sicurezza del Patrimonio Culturale. Nuove metodologie e tecnologie operative-Zaccara Editore. Lagonegro (PZ) 384:27–52

Simmons JE, Muñoz-Saba Y (eds) (2005) Cuidado, manejo y conservación de las colecciones biológicas. Univesidad Nacional de Colombia, Bogotá

Sondi I, Salopek-Sondi B (2004) Silver nanoparticles as antimicrobial agent: a case study on E. coli as a model for Gram-negative bacteria. J Colloid Interface Sci 275(1):177–182

Stankovič J, Cílek V, Bruthans J, Gaál Ľ, Kovács Á, Rozložník M et al (2005) Krásnohorská jaskyňa Buzgó. Speleoklub MINOTAURUS, Rožňava, pp 1–150
Steiger M, Charola AE, Sterflinger K (2011) Weathering and deterioration. In: Stone in architecture. Springer, Berlin, Heidelberg, pp 227–316
Stepien P, Kozlowski R, Tokarz M (1970) Gypstop-colloidal silica for protective coating of porous building materials: practical experience at the Wawel Castle, Cracow, Poland. WIT Trans Built Environ 4
Sterflinger K (2000) Fungi as geologic agents. Geomicrobiol J 17(2):97–124
Sterflinger K (2010) Fungi: their role in deterioration of cultural heritage. Fungal Biol Rev 24 (1–2):47–55
Strzelczyk AB (2004) Observations on aesthetic and structural changes induced in Polish historic objects by microorganisms. Int Biodeterior Biodegrad 53(3):151–156
Strzelczyk AB, Krumbein WE, Majewska L (1989) Öber die Zerstörung historischer Ledereinbände durch Mikroororganismen. Internationale Leder-und Pergamenttagung, vom 8:287–300
Stulik D, Miller D, Khanjian H, Carlson J, Khandekar N, Wolbers R (2004) Solvent gels for the cleaning of works of art: the residue question. Getty Publications, Los Angeles
Szostak-Kotowa J (2004) Biodeterioration of textiles. Int Biodeterior Biodegrad 53(3):165–170
Tang ZX, Fang XJ, Zhang ZL, Zhou T, Zhang XY, Shi LE (2012) Nanosize MgO as antibacterial agent: preparation and characteristics. Braz J Chem Eng 29(4):775–781
Teetor-Barsch GH, Roberts DW (1983) Entomogenous *Fusarium* species. Mycopathologia 84 (1):3–16
Tiano P (2016) Biodeterioration of stone monuments a worldwide issue. Open Conf Proc J 7(1):29
Tiano P, Gargani G (1981) Controlli microbiologici su alcuni affreschi fiorentini. In Atti del convegno sul restauro delle opere d'arte. Firenze, 2–7 Novembre 1976, pp 341–358
Tomšič B, Klemenčič D, Simončič B, Orel B (2011) Influence of antimicrobial finishes on the biodeterioration of cotton and cotton/polyester fabrics: leaching versus bio-barrier formation. Polym Degrad Stab 96(7):1286–1296
UNESCO (2013) Patrimonio Indicadores UNESCO de cultura para el Desarrollo. Available from: Vaillant CM. Materiales constituyentes de las colecciones de archivos y bibliotecas. In: Biodetrioro del patrimonio histórico documental: Alternativas para su erradicación y control. Río de Janeiro: Ed: Mast/FCRB
Vaillant CM (2013) Biodeterioro del Patrimonio Histórico Documental: Alterantivas para su erradicación y Control. Ed. MAST/ FCRB, Río de Janeiro
Valdés-Pérez O, Borrego-Alonso S, Vivar-González I, Anaya-Villalpanda M, Molina-Veloso A (2016) Actividad antifúngica del aceite esencial de clavo de olor en el control del biodeterioro fúngico de documentos. Revista CENIC. Cienc Biol 47(2):78–85
Valentin N (2003) Microbial contamination in museum collections: organic materials. Mol Biol Cult Herit: 85–91
Vásquez Ponce PX (2013) Caracterización y uso del extracto de albahaca como fungicida en bienes patrimoniales maderosos de Quito DM (Bachelor's thesis, Quito: UCE)
Videla HA, Guiamet PS, de Saravia SGG (2003) Biodeterioro de materiales estructurales de sitios arqueológicos de la civilización maya. Revista del Museo de la Plata 44:1–11
Vollrath F (1999) Biology of spider silk. Int J Biol Macromol 24(2–3):81–88
Wainwright M, Ali TA, Barakah F (1993) A review of the role of oligotrophic micro-organisms in biodeterioration. Int Biodeterior Biodegrad 31(1):1–13
Walentowska J, Foksowicz-Flaczyk J (2013) Thyme essential oil for antimicrobial protection of natural textiles. Int Biodeterior Biodegrad 84:407–411
Warscheid T (2000) Integrated concepts for the protection of cultural artifacts against biodeterioration. In: Of microbes and art. Springer, Boston, pp 185–201
Went FW (1969) Fungi associated with stalactite growth. Science 166(3903):385–386

Wheeler G, Mendez-Vivar J, Fleming S (2003) The use of modified Zr-nPropoxide in the consolidation of calcite: a preliminary study focused into the conservation of cultural heritage. J Sol-Gel Sci Technol 26(1–3):1233–1237

Wolbers, R. (2000). Cleaning painted surfaces: aqueous methods

Xie W (1991) Identification and prevention of mould on the frescoes from Chinese Tong tombs. In Proceedings of the EEC China workshop on preservation of cultural heritages. Xian, Shaanxi, PR Of China, September 25–30, 1991, pp 431–438

Zammit G, Sánchez-Moral S, Albertano P (2011) Bacterially mediated mineralisation processes lead to biodeterioration of artworks in Maltese catacombs. Sci Total Environ 409 (14):2773–2782

Microorganisms and Their Enzymes as Biorestoration Agents

5

Chanda Parulekar-Berde, Sachin S. Ghoble, Sagar P. Salvi, and Vikrant B. Berde

Abstract

Cultural heritage (CH) includes art forms such as paintings, frescoes, stoneworks, etc. These are exposed to the combined effects of environmental factors as well as biological activities. The presence of an organic matter in or on the CH such as paintings, frescoes and statues renders them prone to microbial growth. The organic matter used for restoration, for example, glue on paintings and the chemical composition of the art form, serves as a source of carbon and energy, thus allowing the growth of microorganisms. This further degrades the surface of the CH. Consequences of these activities are seen in the form of damaged and deteriorating art forms, which sometimes are totally destroyed. Many methods are available for the cleaning and restoration of the damaged cultural heritage artwork. Use of microorganisms and their enzymes, i.e. biocleaning, amongst these, is the most effective and advantageous. These artistic pathologies can be treated with the help of viable microbial cultures as well as the enzymes produced by microorganisms. Selective isolation of non-pathogenic organic matter degraders and efficient enzyme producers is the starting point in the biorestoration process, followed by the purification of enzymes and their application for biocleaning. This chapter gives a glimpse of the new revolution in the restoration of cultural art forms using enzymes.

Keywords

Cultural heritage · Monuments · Paintings · Biocleaning · Enzymes · Microorganisms

C. Parulekar-Berde (✉) · S. P. Salvi
Department of Microbiology, Gogate Jogalekar College, Ratnagiri, Maharashtra, India

S. S. Ghoble · V. B. Berde
Department of Zoology, Arts, Commerce and Science College, Lanja, Maharashtra, India

A. N. Yadav et al. (eds.), *Microbial Biotechnology Approaches to Monuments of Cultural Heritage*, https://doi.org/10.1007/978-981-15-3401-0_5

5.1 Introduction

Cultural heritage includes various types of artworks such as monuments made of stone, paintings, frescoes, etc. These art forms are not protected the way they should be. As a result of being exposed to biotic and abiotic factors, they undergo deterioration and degradation and eventually lose their aesthetic value. Most significant abiotic factors include light intensity; air pollutants; humidity, temperature, solubility and porosity of the substratum; mechanical strength and nature of the material of the artwork; and chemical nature of the material of CH. Eventually with time, the cultural heritage gets damaged, and changes are seen in the form of discolouration, deposition of dust, black crust formation, blackening, sulphation, physical damage due to growth of filamentous fungi, etc. (Ranalli et al. 1997; Tiano et al. 2006). Depending on the material of the artform, its physical properties differ and so also their susceptibility to deterioration (Saiz-Jimenez 1995; Webster and May 2006). Biodeterioration plays a major role in the damage to cultural heritage. Biodeterioration is the modification of organic and inorganic constituents brought about by the growth and metabolism of microorganisms. The growth of microorganisms is initiated due to the presence of residual compounds either added or present in the art form.

Though responsible for deterioration, microorganisms are also used in the rectification of the damaged art forms. Microorganisms may be applied as a whole or its specific enzyme production ability may be utilized. Biocleaning technologies have been applied to CH monuments like the Camposanto Monumentale Cemetery in Pisa, Italy (Lustrato et al. 2012); the Santos Juanes Church in Valencia, Spain (Bosch-Roig et al. 2013a, b); the Duomo di Milano Cathedral in Milan, Italy (Cappitelli et al. 2006, 2007); the Matera Cathedral, Matera, Italy (May et al. 2008; Alfano et al. 2011); and the Duomo di Firenze Cathedral in Florence, Italy (Gioventù et al. 2011).

5.2 Biodeterioration of Cultural Heritage

Biodeterioration is actually considered as the natural recycling process of degradation of organic and inorganic matters, which is seen happening in any type of material, such as stone, wood, metal, paper and parchment. It involves mostly all artwork including monuments, statues, paintings on canvas and frescoes. Again, the biodeterioration process is strongly influenced by climatic and environmental factors, e.g. extreme temperature, high humidity, low aeration, atmospheric pollution and direct exposure to light and air. All these factors influence the growth of microorganisms, also referred to as biodeteriogens, on the art forms.

5.2.1 Abiotic Factors

Abiotic factors, both physical and chemical, are responsible for the process of biodeterioration of artwork. Environmental and climatic factors such as temperature, humidity, rain, sunlight and organic and inorganic pollutants act on the exposed artwork, making conditions favourable for the growth of microorganisms.

5.2.2 Biotic Factors

The environment can influence the colonization of microbes in a particular habitat. Regions having hot humid climate offer positive environmental conditions for microbial growth. These monuments get degraded and damaged following the growth of microorganisms that form a biofilm over the monument surface, thus eventually decreasing its aesthetic value (Gaylarde et al. 2012; Ortega-Calvo et al. 1991).

Various organic and inorganic constituents present in artworks are responsible for the optimal environment required for the growth and proliferation of microorganisms. The nutritional groups in which the organisms are divided into are based upon the type of constituents utilized by the organisms for metabolism (Caneva et al. 1991). Thus, the chemolithotrophs (sulphur-oxidizing and sulphur-nitrifying bacteria) are included in the autotrophic microorganisms, and photoautotrophic bacteria include the photosynthetic algae and cyanobacteria. Both types of microorganisms are found on artwork like monuments, statues, etc., depending on the type of organic or inorganic materials available (Caneva et al. 1991; Ranalli et al. 2009). The algae and cyanobacteria grow on the art form where sunlight and moisture are available. Hence, these are the primary colonizers. The secondary colonizers utilize the organic substances released by the lysis of primary colonizers or produced by them. The secondary colonizers include the heterotrophs, i.e. bacteria and fungi (Nuhoglu et al. 2006; Ranalli et al. 2009).

The bacterial strains, depending upon the organic matter available, colonize and start the deterioration process. The bacterial genera *Flavobacterium*, *Pseudomonas*, *Alcaligenes*, *Bacillus* and *Staphylococcus* are commonly found to be associated with spoilage of paintings (Capodicasa et al. 2010; López-Miras et al. 2013a, b; Pavić et al. 2015). Fungal growth follows after bacteria colonization, and the commonly found fungal species are *Aspergillus*, *Penicillium*, *Cladosporium* and *Alternaria* (Gorbushina et al. 2004; Caselli et al. 2018). The degradation of artform by microorganisms is thus a combined interaction of number of species of microbes, having degradative enzymes that have important role in the process. Fungal growth caused a physical damage due to the penetration of mycelia deep into the artwork. Even monuments made of stone may get destroyed and damaged as a result of fungal invasion. On paintings, the growth of bacteria and fungi leaves patches of colouration, resulting in paint discolouration, as a result of their metabolic activities. Thus, microbial communities colonizing the art forms and their interactive activities result in the deterioration of the art forms. According to McNamara and Mitchell

(2005), spoilage and deterioration of several CH materials are results of colonization and associative activities in the biofilm formed on art form. An example that showed the prevalence and involvement of microbial growth in deterioration was the study of the growth of autotrophic microbes on marble statue in the Boboli Gardens of Florence (Italy). Cyanobacteria and *Coccomyxa* sp. were observed to colonize the surface, followed by a thin green biofilm of algae in a year of restoration (Lamenti et al. (2000)).

5.3 Organic Substances Produced by Microorganisms Damaging the Cultural Heritage

5.3.1 Organic Acids

Microorganisms produce organic acids in order to make available the components present in the art forms for their growth. This process is called biomineralization and involves the solubilization of components by salt formation and complexation. The various acids produced by microorganisms for the purpose are the citric, lactic, fumaric, itaconic, oxalic, malic, gluconic and sulphuric acids (Table 5.1). De la Rosa-García et al. (2011) reported the production of organic acids by fungi growing on the surface of limestones of Mayan buildings in Yucatan, Mexico. In their study, they showed the formation of calcium oxalate crystals from calcite by the action of fungal organic acids. Some organic acids play multiple roles, for example, oxalic acid helps in the dissolution of siliceous rock as well as forms malonate or calcium oxalate films that protect the calcareous rock surface.

5.3.2 Exopolysaccharide (EPS)

Microorganisms growing in the form of biofilms produce exopolysaccharides that help in the formation and further development of the biofilms (Nwodo et al. 2012). The EPS either may be attached onto the cell forming a capsule or may be released in the medium. It helps in trapping nutrients and organisms. The accumulated nutrients help to tide over nutrient-limiting conditions (Wolfaardt et al. 1998). Most microorganisms form EPS when the concentration of sugars in the media or carbon source is in excess (Petry et al. 2000). Carbohydrates are a predominant component of EPS (Nwodo et al. 2012). Along with some polysaccharides, EPS also contains nucleic acids, lipids, proteins, humic substances and non-carbohydrate constituents like acetate, pyruvate and succinate (Mayer et al. 1999).

5.3.3 Extracellular Enzymes

The microorganisms growing on the art form either in the form of biofilm or singly, produce enzymes that help in the degradation of CH material such as enzyme

Table 5.1 Various organic acids produced by microorganisms

Organic acid	Microorganisms	References
Citric acid	*Penicillium janthinellum* and the species of *Aspergillus*, namely, *A. foetidus*, *A. awamori*, *A. wentii*, *A. carbonarius*, *A. fonsecaeus*, *A. phoenicis*, *Candida oleophila*, *C. guilliermondii*, *Saccaromicopsis lipolytica*, *Hansenula anomala*, *Candida parapsilosis*, *Candida tropicalis*, *C. citroformans*, *Yarrowia lipolytica*, *Bacillus licheniformis*, *Arthrobacter paraffinens* and *Corynebacterium* sp.	Khan et al. (2017), Liaud et al. (2014), Li and Punt (2013), Angumeenal and Venkappayya (2013), Papagianni (2007), Liu et al. (2007), Carta et al. (1999), Kuenz et al. (2012), and Kawamura et al. (1981)
Fumaric acid	*Rhizopus nigricans*, *Circinella*, *Mucor*, *Cunninghamella* and *Rhizopus* species	
Itaconic acid	*Candida* sp., *Ustilago zeae*, *Rhodotorula* sp., *Candida* mutant, *Aspergillus* TN-484-M1 and *Aspergillus terreus terreus* SKR1	
Malic acid	*A. niger*, *A. oryzae*, *A. flavus*, *S. commune*, *Penicillium* sp. K0, *Z. rouxii*, *A. pullulans* ZD-3, *A. pullulans* ZX-10, *P. rubens*, *P. viticola*, *Aureobasidium* sp. P6, *S. cerevisiae*, *A. oryzae* NRRL	
Propionic acid	*Propionibacterium* sp. like *P. acidipropionici*, *Veillonella parvula*, *V. alcalescens*, *Selenomonas ruminantium*, *S. sputigena*, *P. freudenreichii*, *P. shermanii*, *Propionigenum modestum*	
Oxalic acid	*Aspergillus niger*	

tyrosinase in the process of biodeterioration. In the case of biofilms, the enzymes produced by the microbes get accumulated and also are protected from desiccation and radiation (Sabater et al. 2016). The enzymes produced may include β-glucosidases, β-mannosidases, cellulases, amylases, xylanases, etc. The enzymes are responsible for the deterioration and discolouration of cultural monuments, for example, the blackening of stone crust due to growth of microorganisms and action of enzymes. Warscheid and Braams (2000) report the role of enzymes released from microorganisms that result in melanin formation and thus the blackening observed. Melanin production is beneficial to the microorganisms as it protects them from UV rays, desiccation, temperature changes and any hydrolytic enzymes secreted.

5.4 Bioremediation of CH Using Microorganisms

Bioremediation uses bacteria that have the ability to produce enzymes which biocalcify, reduce sulphate and reduce nitrate and have hydrolysing activities. The enzymes are preferably lipases, proteases and carbohydrases. Biocalcifying bacteria are usually used for stone consolidation (Tiano et al. 1999). The sulphate reducers and nitrate reducers are used for conversion of sulphate and nitrate salts included in the mineral matrix, respectively. Sulphate-reducing bacteria reduce sulphate to gaseous hydrogen sulphide, while nitrate-reducing bacteria reduce nitrates to gaseous nitrogen or nitrous oxide. The hydrolytic enzymes play a major role in the removal of organic substances (Saiz-Jimenez 1997; Wolbers 2000; Fernandes 2006).

5.4.1 Microorganisms Used for Bioremediation of Cultural Heritage

The application of the metabolic capabilities of the microbes has been defined as microbial resource management (MRM) (Verstraete et al. 2007), and the use of bacterial cells in biorestoration is termed as biocleaning (Ranalli et al. 2000). For the isolation of efficient biocleaning bacteria, it is recommended to use varnish, paint or industrial wastewater of polymer industry and sludge as a source (Arutchelvan et al. 2005; Chen et al. 2007). A number of workers have reported the isolation of appropriate microorganisms, having the ability to remove undesirable substances, using this methodology (Ranalli et al. 2000; Troiano et al. 2014).

Ranalli et al. (1996) were the first to apply nitrate-reducing bacteria (NRB) for the bioremediation, i.e. biocleaning of Vicenza stone art that was damaged due to nitrates. A strain of *Pseudomonas stutzeri*, delivered onto the art form in sepiolite, was used for a purpose. *Halomonas campaniensis* spp. have been used for the biocleaning of nitrate crusts on stone surfaces (Ranalli et al. 2019). The growth of the species is supported by the crust composition which includes sulphate deposits, nitrate salts and other compounds such as carbonates, apatite and protein traces. Hence, removal of the crust is very important.

Yeast *Candida cylindracea* lipase was used for the removal of Paraloid B-72 from two paintings (Bellucci et al. 1999). Similarly, there is another report of the application of a yeast culture isolated from bronze statue treated with Incralac (copolymer of ethyl methacrylate and butyl acrylate) for degradation of the same coating (McNamara et al. 2004). Yeast is suitable for the biocleaning process, as it is safe for the environment and user with low toxic metabolite production (Saleem et al. 2008).

Other microbial species that have an efficient use in the biocleaning processes are the sulphate reducers, *Desulfovibrio desulfuricans* and *Desulfovibrio vulgaris*, and nitrate-reducing bacteria such as *Pseudomonas stutzeri*, *Pseudomonas denitrificans*, *Pseudomonas aeruginosa*, *Pseudomonas pseudoalcaligenes* and *Paracoccus denitrificans* (Heselmeyer et al. 1991; Gauri et al. 1989, 1992; Ranalli et al. 1996, 2000, 2003, 2005; Ranalli and Sorlini 2008; Cappitelli et al. 2005, 2006, 2007; Polo et al. 2010; Alfano et al. 2011).

5.4.2 Microbial Removal of Organic Substances Damaging the Cultural Heritage

The major causes of deterioration of art forms are the accumulation of organic matter in/on CH artwork and the consequent growth of microorganisms. Thus, removal of the organic substances from the art forms is a major step in the restoration of CH. The various organic substances found are casein, eggyolk, oil and animal fat, which are added to art forms during restoration processes or may be added during art form creation. Removal of these components is difficult or more damaging, with physical and chemical techniques. The removal technique requires a shorter application time and, along with effectiveness, should be easy to apply (Ranalli et al. 2019). But the modern cleaning techniques, such as the enzymes, surfactants and solubilizing agents, are a little less efficient when it comes to restoration of frescoes (Makes 1988; Bellucci and Cremonesi 1994; Bonomi 1994; Wolbers 2007). Thus, the use of whole cells of microorganisms is the best methodology. The bioremoval of animal glue from a Spinello Aretino fresco in the Camposanto Monumental Cemetery in Pisa using *Pseudomonas stutzeri* is an example of positive results obtained using microorganisms (Ranalli et al. 2009). *P. stutzeri* has also been used for the removal of organic substances from wall paintings by a number of researchers (Ranalli and Sorlini 2008; Antonioli et al. 2005; Ranalli et al. 2005; Sorlini and Cappitelli 2008; Polo et al. 2010).

There are many reports of use of *Pseudomonas stutzeri* for the removal of organic matters from wall paintings (Ranalli et al. 2005; Antonioli et al. 2005; Ranalli et al. 2003; Lustrato et al. 2012; Bosch-Roig et al. 2013; Ranalli et al. 2018). The glue applied between painted frescoes and the adhering gauze for detaching frescoes from walls during the Second World War are examples of art form containing organic substances. In situ tests showed the use of viable cells is a very versatile approach as compared to chemical or mechanical methods (Ranalli et al. 2019).

Biocleaning methodologies of cultural monuments are aimed at the removal of organic matter on the surface of the monuments and art forms. Bioremediation has been carried out on the external walls of the Matera Cathedral in Italy and the wall paintings placed in the lunettes of the central vault of the Santos Juanes Church in Valencia, Spain (Alfano et al. 2011; Roig et al. 2013). The external wall of the cathedral of Matera was treated with *Pseudomonas pseudoalcaligenes* cells entrapped in a Carbogel carrier and then covered with a PET film. *Pseudomonas stutzeri* DSMZ 5190 was applied on wall painting surfaces of the Santos Juanes Church in Valencia by means of Japanese paper and agar carrier layers. Infrared heat lamps were used during the treatment to ensure the correct metabolic activity of the bacteria.

The success of biotreatments with viable microorganisms can be attributed to the high versatility of the bacteria and their wide range of enzymatic activities, both constitutive and inducible. Their use is more effective as compared to the treatment with enzymes such as proteases and collagenases, and in general the use of individual enzymes is highly specific and limited, resulting also in less convenience, compared to the cost of bacterial biorestoration (Ranalli et al. 2005, 2019).

A biocleaning approach has recently been evaluated also for the removal of graffiti paint, a very common cause of disfigurement of monuments, buildings and other urban structures. Sanmartín et al. investigated an efficient biocleaning protocol for in situ application, reducing time of treatment and using a specific culture medium for the adaptation of bacteria to the target substrate (Giacomucci et al. 2012; Sanmartín et al. 2015; Sanmartín and Bosch-Roig 2019).

Even though the results of microbial treatment methods are very promising, the traditional methods are preferred. The reason being the concern that microorganisms may be infectious and the safety of workers or users is in question. Hence, it follows that demonstrating the safety of biocleaning treatments is an essential step required for the adoption of biorestoration methods for conservation of items of historic and cultural importance.

5.5 Enzymes as Agents of CH Remediation

Biological methods that use microorganisms and enzymes as biological cleaning agents in the "biorestoration" of artworks are becoming attractive alternatives to the mechanical and chemical methods. They offer significant advantages in terms of soft intervention on the works themselves, lack of health risks for conservator-restorers and also guaranteeing environmental safety (Saiz-Jimenez 1997; Cremonesi 2002; Ranalli et al. 2005; Valentini et al. 2010). Under optimal controlled conditions, biological methods reproduce the same processes that occur in nature (Boquet et al. 1973; Atlas et al. 1988; Ferrer et al. 1988; Heselmeyer et al. 1991; Tiano et al. 1996; Castanier et al. 2000; Maier et al. 2000; Rodriguez-Navarro et al. 2000; Zanardini et al. 2000; Ranalli et al. 2003; Biavati and Sorlini 2008). In fact, bacteria are known to produce not only constitutive but also inducible enzymes whose synthesis takes place only in the presence of a specific substrate. These enzymes can attack and degrade different types of molecules only when the bacterial cells are exposed to them, creating a regulatory effect.

Based on previous lab trials (Ranalli et al. 2003, 2005), a suspension of *Pseudomonas stutzeri* (DSMZ 5190) was used for the treatment of the fresco implanted in hydrophilic cotton strips. Antonioli et al. (2005) discovered that *P. stutzeri* DSMZ 5190 produces caseinolytic and collagenase activities, which had the ability to degrade the main animal glue constituents.

According to Decoux (2002), several commercial enzymes have the ability to remove adhesives. There are reports on hydrolytic enzymes acting on animal glues (Pangallo et al. 2012; Jeszeová et al. 2018; Kraková et al. 2018). The mechanical scraping of artwork can damage it and can seriously affect the integrity of the sheets (Vávrová and Součková 2017). The use of enzymes should reduce the amount of scraping needed, thereby avoiding damaging the newspapers.

Biocleaning protocols using new hydrolases isolated from marine organisms are being worked on where they make use of hydrolytic enzymes (Ranalli et al. 2005; Hamed 2012). There are reports of two batches of hydrolase, with collagenolytic-proteolytic activity and the protease isolated from vertebrate and invertebrate marine organisms (Salamone et al. 2012; Kembhavi et al. 1993). These enzymes are active in a range of temperatures between 4 and 37 °C, with the activity of removal of glue, at normal room temperatures.

The cleaning tests showed reliable results, without the heating of the enzyme solutions or of the surfaces on which they were applied (Palla et al. 2012; Palla et al. 2013). This particular feature makes these proteases more appropriate than others, usually active at higher temperatures. In the year 1970, the first biocleaning attempts on paper-based artefacts and polychrome canvases using enzymes were reported. Later, Wendelbo reported the treatment of pages of a book pasted together with animal glue with trypsin (Wendelbo and Fosse 1970; Wendelbo 1976). A combination of two enzymes, amylase and protease, was used by Segal and Cooper for the removal of adhesives composed of starch and protein (Segal and Cooper 1977). The glue paste from the back of an oil painting was removed using a mixture of the two enzymes (Sampá and LuppiMosca 1989). Later in 1988, there was another similar report of use of mixture of enzymes for removal of a protein/oily binder from a painted surface (Makes 1988).

Apart from proteases and amylases, lipases have also been applied in different restoration procedures. Lipases represent a valid alternative to conventional methods for removing aged siccative oil, as they represent safer working conditions with regard to both the operator's health and the integrity of the artwork. Workers have isolated, purified and worked with lipolytic enzymes as well as proteases of marine nature, for bioremediation of artwork especially paper. Paraloid B-72, an acrylic resin, was removed from two paintings using lipases (Vokic and Berovic 2005; Bellucci et al. 1999). The undesirable colouration is caused by growth of bacteria, such as *Serratia*, on artwork. Removal of this colouration by mechanical methods can result in damage to the art form. Fungal enzyme laccase has also been used in the biocleaning process of red stain removal from marble statue of Isamu Noguchi's Slide Mantra (Konkol et al. 2009). The enzyme was obtained from the fungal culture *Trametes versicolor*.

The technological biocleaning methodology is also based on the use of combination of viable microbial cells and hydrolytic enzymes so as to ascertain the complete removal of unwanted layers on artwork surfaces (Ranalli et al. 2005; Bosch-Roig et al. 2013b). There are reports of the biocleaning of one of the frescoes in the Pisa monumental cemetery by using *Pseudomonas* viable bacterial cells. The organic matter was removed by the bacteria, while the residual glue was removed by protease (type XIX) enzyme solution (Ranalli et al. 2005).

Hydrolytic enzymes signify a very helpful means in the remediation of a diversity of artworks, particularly to eliminate dirts, adhesives, other organic residues from paintings, wooden and stone artworks, mural paintings and papers (Ranalli et al. 2005; Hamed 2012; Valentini et al. 2012; Barbabietola et al. 2016). Various hydrolases with effective biocleaning properties are commercially available. The source of the enzyme is either from animal, plant, bacterial, or fungal tissues (Ranalli et al. 2005; Palla et al. 2013). For example, pepsin, trypsin and protease can be extracted from both animal tissues (pancreas, stomach) and microbial cells (*Bacillus*, *Aspergillus*). Based on certain aspects, the selection of enzymes can be done, such as the material to be removed, enzyme specificity, physico-chemical parameters for enzyme activity and safety of the user. Commercial enzymes require an ambient temperature for activity, i.e. ≥37 °C. This limits its use on a wider scale. Also substances such as salts, metal ions, pigments or other molecules on the artwork can act as inhibitors of the enzyme activity (Bellucci and Cremonesi 1994). After biocleaning of the artwork for the removal of unwanted layers of material, the artwork should be checked for residue presence to estimate possible chromatic alterations in the surface (Pruteanu et al. 2014; Micheli et al. 2016; Hrdlickova Kuckova et al. 2014; Bosch-Roig et al. 2013a, b; Palla et al. 2016).

5.6 Prevention and Control of Biodeterioration of CH

Prevention and recovery are the two most important aspects of the conservation of cultural heritage. The methods that prevent and minimize deterioration due to microorganisms are called prevention methods or indirect methods. These methods are aimed at reducing or inhibiting microbial growth and controlling ambient parameters. For example, growth in places like museums, churches and indoor environments can be monitored for humidity, lighting, aeration and temperature. In order to avoid the growth of microorganisms, these parameters need to be controlled. Apart from controlling these physico-chemical parameters, cleaning methods are also important in the elimination of substances that will contribute to the microbial growth, for example, removal of dirt and dust which are sources of nutrition for many microorganisms and are also sources of inoculum as they carry spores and microorganisms. Further, Sterflinger and Piñar (2013) have suggested that it is very important to undertake frequent cleaning as dust harbours a large quantity of bacteria and fungi, which can eventually colonize the cultural artefacts.

Prevention and control of outdoor conditions differ from indoor conditions. The conditions listed above are not applicable for outdoor CH. It is very difficult or impossible to maintain physico-chemical parameters outdoors in exposed environment. However, microbial growth can be prevented by routine maintenance and periodic monitoring. Here, we can make use of remedial methods or the direct methods. The biodeteriogens, i.e. organisms involved in biodeterioration, themselves are eradicated. The methods used may vary from mechanical and physical to chemical. The latest and the most effective is the use of microorganisms and their enzymes.

5.7 Conclusion and Future Prospects

The biocleaning methods are however still not used routinely in CH restoration. Though it seems to be most effective, there are a number of issues that need to be addressed. The methods vary depending on the material of artwork or the type of matter to be removed; the application methods need to be more developed or standardized and so on. The potentiality of bioremediation processes, using microorganisms, for the restoration of artwork especially in the removal of organic matter has been proved. However, with whole-cell use, the question of human risk arises. The advantages of using enzymes (or living bacteria) as an alternative to traditional methods as well as to use of whole live cells lie in the fact that the enzymes are more specific and highly selective in their activities and are non-invasive. The enzymes act on their target molecules only and do not attack other molecules. Thus, purified enzymes represent a valid alternative to conventional methods for removal of organic substances from CH, in a safe way, as regards both the operator's health and the artwork's integrity. Moreover, purified enzymes with ambient temperatures below 30 °C, such as those that can be isolated from marine organisms, are extremely useful in the restoration of historic-artistic works of art. They are important for testing potential enzymatic mixtures and for optimizing specific biocleaning procedures for cultural heritage items.

Acknowledgements The authors are thankful to their respective institutions for the encouragement and support.

References

Alfano G et al (2011) The bioremoval of nitrate and sulfate alterations on artistic stonework: the case-study of Matera cathedral after six years from the treatment. Int Biodeterior Biodegradation 65:1004–1011

Angumeenal AR, Venkappayya D (2013) An overview of citric acid production. LWT Food Sci Technol 50:367–370

Antonioli P, Zapparoli G, Abbruscato P, Sorlini C, Ranalli G, Righetti PG (2005) Art-loving bugs: the resurrection of Spinello Aretino from Pisa's cemetery. Proteomics 5:2453–2459

Arutchelvan V, Kanakasabai V, Nagarajan S, Muralikrishnan V (2005) Isolation and identification of novel high strength phenol degrading bacterial strains from phenol-formaldehyde resin manufacturing industrial wastewater. J Hazard Mater 127:238–243

Atlas RM, Chowdhury AN, Gauri KL (1988) Microbial calcification of gypsum-rock and sulfated marble. Stud Conserv 33(3):149–153

Barbabietola N, Tasso F, Alisi C, Marconi P, Perito B, Pasquariello G, Sprocati AR (2016) A safe microbe-based procedure for a gentle removal of aged animal glues from ancient paper. Int Biodeterior Biodegradation 109:53–60

Bellucci R, Cremonesi P (1994) L'usodeglienziminellaconservazione e nelrestaurodeidipinti. The use of enzymes in the conservation and restoration of paintings. Kermes arte e tecnicadelrestauro 21:45–64

Bellucci R, Cremonesi P, Pignagnoli G (1999) A preliminary note on the use of enzymes in conservation. The removal of aged acrylic resin coatings with lipase. Stud Conserv 44:278–281

Biavati B, Sorlini C (2008) Microbiologiaagroambientale. Ambrosiana, Milano, p 684

Bonomi R (1994) Utilizzodeglienzimi per ilrestauro di unascultura in terracotta policroma. OPD Restauro 6:101–107

Boquet E, Boronat A, Ramos-Cormenzana A (1973) Production of calcite (calcium carbonate) crystals by soil bacteria is a general phenomenon. Nature 246:527–529

Bosch-Roig P, Regidor-Ros JL, Montes-Estellés R (2013a) Biocleaning of nitrate alterations on wall paintings by *Pseudomonas stutzeri*. Int Biodeterior Biodegrad 84:266–274

Bosch-Roig P, Regidor-Ros JL, Soriano-Sancho P, Montes-Estellés RM (2013b) Biocleaning of animal glue on wall paintings by *Pseudomonas stutzeri*. Chin Oggi/Chem Today 31(1):50–53

Caneva G, Nugari MP, Salvadori O (1991) Environmental factors in biodeterioration. In: Caneva G, Nugari MP, Salvadori O (eds) Biology in the conservation of works of art. ICCROM—SintesiGraficas.r.l., Rome, pp 3–24
Capodicasa S, Fedi S, Porcelli AM, Zannoni D (2010) The microbial community dwelling on a biodeteriorated 16th century painting. Int Biodeterior Biodegrad 64:727–733
Cappitelli F, Zanardini E, Toniolo L, Abbruscato P, Ranalli G, Sorlini C (2005) Bioconservation of the marble base of the Pietà Rondanini by Michelangelo Buonarroti. J Appl Microbiol 7:06675
Cappitelli F, Principi P, Sorlini C (2006) Biodeterioration of modern materials in contemporary collections: can biotechnology help? Trends Biotechnol 24(8):350–354
Cappitelli F, Toniolo L, Sansonetti A, Gulotta D, Ranalli G, Zanardini E, Sorlini C (2007) Advantages of using microbial technology over traditional chemical technology in removal of black crusts from stone surfaces of historical monuments. Appl Environ Microbiol 73:5671–5675
Carta FS, Soccol CR, Ramos LP, Fontana JD (1999) Production of fumaric acid by fermentation of enzymatic hydrolysates derived from cassava bagasse. Bioresour Technol 68:23–28
Caselli E, Pancaldi S, Baldisserotto C, Petrucci F, Impallaria A, Volpe L, D'Accolti M, Soritti I, Coccagna M, Sassu G et al (2018) Characterization of biodegradation in a 17th century easel painting and potential for a biological approach. PLoS One 13:e0207630
Castanier S, Le Metayer-Levrel G, Orial G, Loubiere JF, Perthuisot JP (2000) Carbonatogenesis and applications to preservation and restoration of historic property. In: Ciferri O, Tiano P, Mastromei G (eds) Of microbes and art. Kluwer Academic/Plenum Publishers, Amsterdam, pp 203–218
Chen J, Zhang Y, Du GC, Hua ZZ, Zhu Y (2007) Biodegradation of polyvinyl alcohol by a mixed microbial culture. Enzym Microb Technol 40(7):1686–1691
Cremonesi P (2002) L'usodeglienziminellapulitura di operepolicrome. Il Prato, Padova
De la Rosa-García C, Ortega-Morales O, Claire Gaylarde C, Beltrán-García M, Quintana-Owen P, Reyes-Estebanez M (2011) Influence of fungi in the weathering of limestone of Mayan monuments. Rev Mex Micol 33:43–51
Decoux S (2002) Enzymes used for adhesive removal in paper conservation: a literature review. J Soc Arch 23:187–195
Fernandes P (2006) Applied microbiology and biotechnology in the conservation of stone cultural heritage materials. Appl Microbiol Biotechnol 73:291–296
Ferrer MR, Quevedo-Sarmiento J, Rivadeneira MA, Bejar V, Delgadoand R, Ramos-Cormenzana A (1988) Calcium carbonate precipitation by two groups of moderately halophilic microorganisms at different temperatures and salt concentrations. Curr Microbiol 17:221–227
Gauri LK, Chowdhury AN, Kulshreshtha NP, Punuru AR (1989) The sulfation of marble and the treatment of gypsum crusts. Stud Conserv 34:201–206
Gauri LK, Parks L, Jaynes J, Atlas R (1992) Removal of sulphated crust from marble using sulphate-reducing bacteria, pp 160–165. In: Robin GM (ed) Stone cleaning and the nature, soiling and decay mechanisms of stone. Proceedings of the international conference, 14 to 16 April 1992. Donhead Publishing Ltd., Webster, Edinburgh
Gaylarde CC, Rodrıguez CH, Navarro NY, Ortega MB (2012) Microbial biofilms on the sandstone monuments of the Angkor Wat complex, Cambodia. Curr Microbiol 64:85–92
Giacomucci L, Toja F, Sanmartín P, Toniolo L, Prieto B, Villa F, Cappitelli F (2012) Degradation of nitrocellulose-based paint by *Dedesulfuricans* ATCC 13541. Biodegradation 23:705–716
Gioventù E, Lorenzi PF, Villa F, Sorlini C, Rizzi M, Cagnini A et al (2011) Comparing the bioremoval of black crusts on colored artistic lithotypes of the cathedral of Florence with chemical and laser treatment. Int Biodeterior Biodegradation 65:832–839
Gorbushina AA, Heyrman J, Dornieden T, Gonzalez-Delvalle M, Krumbein WE, Laiz L, Petersen K, Saiz-Jimenez C, Swings J (2004) Bacterial and fungal diversity and biodeterioration problems in mural painting environments of St. Martins church (Greene–Kreiensen, Germany). Int Biodeterior Biodegrad 53:13–24

Hamed SAM (2012) A preliminary study on using enzymes in cleaning archaeological wood. J Archaeol Sci 39(7):2515–2520

Heselmeyer K, Fischer U, Krumbein WE, Warsheid T (1991) Application of *Desulfovibrio vulgaris* for the bioconversion of rock gypsum crusts into calcite. Bioforum 1:89

HrdlickovaKuckova S, CrhovaKrizkova M, Pereira CLC, Hynek R, Lavrova O, Busani T, Branco LC, Sandu ICA (2014) Assessment of green cleaning effectiveness on polychrome surfaces by MALDI-TOF mass spectrometry and microscopic imaging. Microsc Res Tech 77(8):574–585

Jeszeová L, Bauerová-Hlinková V, Baráth P, Puškárová A, Bučková M, Kraková L, Pangallo D (2018) Biochemical and proteomic characterization of the extracellular enzymatic preparation of *Exiguobacterium undae*, suitable for efficient animal glue removal. Appl Microbiol Biotechnol 102:6525–6536

Kawamura D, Furuhashi M, Saito O, Matsui H(1981) Production of itaconic acid by fermentation. Jp Patent .56: 137,893 (To Iwata)

Kembhavi AA, Kulkarni A, Pant A (1993) Salt-tolerant and thermostable alkaline protease from *Bacillus subtilis*. NCIM no.64. Appl Biochem Biotechnol 38:83–92

Khan I, Qayyum S, Maqbool F, Mujaddad-ur-Rehman, Hayat A, Farooqui SM (2017) Microbial organic acids production, biosynthetic mechanism and applications. Indian J Geo Mar Sci 46:2165–2174

Konkol N, Macnamara C, Sembrat J, Rabinowitz M, Mitchell R (2009) Enzymatic decoloration of bacterial pigmented from culturally significant marble. J Cult Herit 10:362–366

Kraková L, Šoltys K, Puškárová A, Bučková M, Jeszeová L, Kucharík M, Budiš J, Orovčík L, Szemes T, Pangallo D (2018) The microbiomes of a XVIII century mummy from the castle of Krásna Hôrka (Slovakia) and its surrounding environment. Environ Microbiol 20:3294–3308

Kuenz A, Gallenmuller Y, Willke T, Vorlop KD (2012) Microbial production of itaconic acid: developing a stable platform for high product concentrations. Appl Microbiol Biotechnol 96 (5):1209–1216

Lamenti G, Tiano P, Tomaselli L (2000) Biodeterioration of ornamental marble statues in the Boboli gardens (Florence, Italy). J Appl Phycol 12:427–433

Li A, Punt P (2013) Industrial production of organic acids by Fungi. In: Gupta V, Schmill M, Mazutti M, Mäki M, Tuohy M (eds) Applied microbiology engineering. Taylor and Francis Group, London, pp 52–74

Liaud N, Giniés C, Navarro D, Fabre N, Crapart S, Gimbert I, Levasseur A, Raouche S, Sigoillot JC (2014) Exploring fungal biodiversity: organic acid production by 66 strains of filamentous fungi. Fungal Biol Biotechnol 1:1

Liu LM, Li Y, Zhu Y, Du GC, Chen J (2007) Redistribution of carbon flux in *Torulopsis glabrata* by altering vitamin and calcium level. Metab Eng 9:21–29

López-Miras MDM, Martín-Sánchez I, Yebra-Rodríguez Á, Romero-Noguera J, Bolívar-Galiano F, Ettenauer J et al (2013a) Contribution of the microbial communities detected on an oil painting on canvas to its biodeterioration. PLoS One 8(11):e80198

López-Miras M, Piñar G, Romero-Noguera J, Bolívar-Galiano FC, Ettenauer J, Sterflinger K et al (2013b) Microbial communities adhering to the obverse and reverse sides of an oil painting on canvas: identification and evaluation of their biodegradative potential. Aerobiologia 29 (2):301–314

Lustrato G, Alfano G, Andreotti A, Colombini MP, Ranalli G (2012) Fast biocleaning of mediaeval frescoes using viable bacterial cells. Int Biodeterior Biodegrad 69:51–61

Maier RM, Pepper IL, Gerba CP (2000) Environmental microbiology. Elsevier, San Diego, p 585

Makes F (1988) Enzymatic consolidation of the portrait of Rudolf II with a multienzyme preparation isolated from Antarctic krill, Acta Universitatis Gothoburgensis – Goteborg Studies on Conservation, 1, Goteborg, Sweden

May E, Webster AM, Inkpen R, Zamarreno D, Kuever J, Rudolph C, Warscheid T, Sorlini C, Cappitelli F, Zanardini E et al (2008) The biobrush project for bioremediation of heritage stone. In: May E, Jones M, Mitchell J (eds) Heritage microbiology and science: microbes, monuments and maritime materials. RSC Publishing, Cambridge, UK, pp 76–93

Mayer C, Moritz R, Kirschner C, Borchard W, Maibaum R, Wingender J, Flemming HC (1999) The role of intermolecular interactions: studies on model systems for bacterial biofilms. Int J Biol Macromol 26:3–16
McNamara CJ, Mitchell R (2005) Microbial deterioration of historic stone. Front Ecol Environ 3:445–451
McNamara CJ, Breuker M, Helms M, Perry TD, Mitchell R (2004) Biodeterioration of Incralac used for the protection of bronze monuments. J Cult Herit 5:361–364
Micheli L, Mazzuca C, Palleschi A, Palleschi G (2016) Development of a diagnostic and cleaning tool for paper artworks: a case of study. Microchem J 126:32–41
Nuhoglu Y, Oguz E, Uslu H, Ozbek A, Ipekoglu B, Ocak I et al (2006) The accelerating effects of the microorganisms on biodeterioration of stone monuments under air pollution and continental-cold climatic conditions in Erzurum, Turkey. Sci Total Environ 364:272–283
Nwodo U, Green E, Okoh A (2012) Bacterial exopolysaccharides: functionality and prospects. Int J Mol Sci 13:14002–14015
Ortega-Calvo JJ, Naturales R, Saiz-Jiminez C (1991) Biodeterioration of building materials by cyanobacteria and algae. Int Biodeterior 28:165–185
Palla F, Cammarata M, Trapani MR, Billeci N, Sebastianelli M, Salamone M, Ghersi G (2012) Removal of old protein layers from surfaces of works of art by new enzymes, 36th International Symposium on the Conservation and Restoration of Cultural Properties, pp 78–79, 5–7 Dicembre, Tokyo, Japan
Palla F, Cammarata M, Trapani MR, Ghersi G, Salamone M, Sebatianelli M (2013) Novel proteases from marine organisms with potential interest in restoration procedure. In: Science and technology for the conservation of cultural heritage. CRC Press, Taylor and Francis Group
Palla F, Barresi G, Giordano A, Schiavone S, Trapani MR, Rotolo V, Parisi MG, Cammarata M (2016) Cold-active molecules for a sustainable preservation and restoration of historic-artistic manufacts. Int J Conserv Sci 7:239–246
Pangallo D, Kraková L, Chovanová K, Šimonovičová A, De Leo F, Urzì C (2012) Analysis and comparison of the microflora isolated from fresco surface and from surrounding air environment through molecular and biodegradative assays. World J Microbiol Biotechnol 28:2015–2027
Papagianni M (2007) Advances in citric acid fermentation by aspergillus niger: biochemical aspects, membrane transport and modeling. Biotechnol Adv 25(3):244–263
Pavić A, Ilić-Tomić T, Pačevski A, Nedeljković T, Vasiljević B, Morić I (2015) Diversity and biodeteriorative potential of bacterial isolates from deteriorated modern combined-technique canvas painting. Int Biodeterior Biodegrad 97:40–50
Petry S, Furlan S, Crepeau MJ, Cerning J, Desmazeaud M (2000) Factors affecting exocellular polysaccharide production by *Lactobacillus delbrueckii* subsp. *bulgaricus* grown in a chemically defined medium. Appl Environ Microbiol 66:3427–3431
Polo A, Cappitelli F, Brusetti L, Principi P, Villa F, Giacomucci L, Ranalli G, Sorlini C (2010) Feasibility of removing surface deposits on stone using biological and chemical remediation methods. Microb Ecol 60(1):1–14
Pruteanu S, Vasilache V, Sandu ICA, Budu AM, Sandu I (2014) Assessment of cleaning effectiveness for new ecological systems on ancient tempera icon by complementary microscopy techniques. Microsc Res Tech 77(12):1060–1070
Ranalli G, Sorlini C (2008) Bioremediation. In: Caneva G, Nugari MP, Salvadori O (eds) Plant biology for cultural heritage. Bioremediation and conservation. The Getty Conservation Institute, Los Angeles, pp 340–346
Ranalli G, Chiavarini M, Guidetti V, Marsala F, Matteini M, Zanardini E, Sorlini C (1996) The use of microorganisms for the removal of nitrates and organic substances on artistic stoneworks. In: Riederer J (ed) Proceedings 8th international congress on deterioration and conservation of stone. Möller, Berlin, pp 1415–1420
Ranalli G, Chiavarini M, Guidetti V, Marsala F, Matteini M, Zanardini E, Sorlini C (1997) The use of microorganisms for the removal of sulphates on artistic stoneworks. Int Biodeterior Biodegrad 40:255–261

Ranalli G, Matteini M, Tosini I, Zanardini E, Sorlini C (2000) Bioremediation of cultural heritage: removal of sulfates, nitrates and organic substances. In: Ciferri O, Tiano P, Mastromei G (eds) Of microbes and art: bioremediation of cultural heritage: removal of sulphates, nitrates and organic substances. Springer, New York

Ranalli G, Belli C, Baracchini C, Caponi G, Pacini P, Zanardini E, Sorlini C (2003) Deterioration and bioremediation of frescoes: a case study. In: Saiz-Jimenez C (ed) Molecular biology and cultural heritage. Balkema Publishers, Lisse, pp 243–246

Ranalli G, Alfano G, Belli C, Lustrato G, Colombini MP, Bonaduce I, Zanardini E, Abbruscato P, Cappitelli F, Sorlini C (2005) Biotechnology applied to cultural heritage: biorestoration of frescoes using viable bacterial cells and enzymes. J Appl Microbiol 98:73–83

Ranalli G, Zanardini E, Sorlini C (2009) Biodeterioration—including cultural heritage. In: Schaechter M (ed) Encyclopedia of microbiology, 3rd edn. Academic Press, Oxford, pp 191–205

Ranalli G, Zanardini E, Andreotti A, Colombini MP, Corti C, Bosch-Roig P, De Nuntiis P, Lustrato G, Mandrioli P, Rampazzi L et al (2018) Hi-tech restoration by two-steps biocleaning process of triumph of death fresco at the Camposanto monumental cemetery (Pisa, Italy). J Appl Microbiol 125:800–812

Ranalli G, Zanardini E, Rampazzi L, Corti C, Andreotti A, Colombini MP, Bosch-Roig P, Lustrato G, Giantomassi C, Zari D et al (2019) Onsite advanced biocleaning system on historical wall paintings using new agar-gauze bacteria gel. J Appl Microbiol 126:1785–1796

Rodriguez-Navarro C, Rodriguez-Gallego M, Ben Chekroun K, Gonzales-Munoz MT (2000) Carbonate production by *Myxococcus xanthus*: a possible application to protect/consolidate calcareous stones. Proceeding of the International Congress Quarry, Laboratory, Monument, Pavia, pp 493–498

Roig PB, Ros JLR, Montes Estellés R (2013) Biocleaning of nitrate alterations on wall paintings by *Pseudomonas stutzeri*. Int Biodeterior Biodegrad 84:266–274

Sabater S, Timoner X, Borrego C, Acuna V (2016) Stream biofilm responses to flow intermittency: from cells to ecosystems. Front Environ Sci 4:14

Saiz-Jimenez C (1995) Deposition of anthropogenic compounds on monuments and their effect on airborne microorganisms. Aerobiologia 11:161–175

Saiz-Jimenez C (1997) Biodeterioration vs biodegradation: the role of microorganisms in the removal of pollutants deposited on to historic buildings. Int Biodeterior Biodegrad 40:225–232

Salamone M, Cuttitta A, Seidita G, Mazzola S, Bertuzzi F, Ricordi C, Ghersi G (2012) Characterization of collanolytic/proteolytic marine enzyme. Chem Eng Trans 27:1–6

Saleem M, Brim H, Hussain S, Arshad M, Leigh MB (2008) Perspectives on microbial cell surface display in bioremediation. Biotechnol Adv 26(2):151–161

Sampá S, LuppiMosca AM (1989) A study of the fungi occurring on 15th century frescoes in Florence, Italy. Int Biodeterior 25(5):343–353

Sanmartín P, Bosch-Roig P (2019) Biocleaning to remove graffiti: a real possibility? Advances towards a complete protocol of action. Coatings 9:104

Sanmartín P, DeAraujo A, Vasanthakumar A, Mitchell R (2015) Feasibility study involving the search for natural strains of microorganisms capable of degrading graffiti from heritage materials. Int Biodeterior Biodegrad 103:186–190

Segal J, Cooper D (1977) The use of enzymes to release adhesives. Pap Conserv 2(1):47–50

Sorlini C, Cappitelli F (2008) The application of viable bacteria for the biocleaning of cultural heritage surfaces. Coalition 15:18e20

Sterflinger K, Piñar G (2013) Microbial deterioration of cultural heritage and works of art – tilting at windmills? Appl Microbiol Biotechnol 97(22):9637–9646

Tiano P, Tosini I, Rizzi M, Tsakoma M (1996) Calcium oxalate decomposing microorganisms: a biological approach to the oxalate patinas elimination. In: Realini M, Toniolo L (eds) Proceedings of the 2nd international symposium: the oxalate films in the conservation of works of art. Editream, Milan, pp 25–27

Tiano P, Biagiotti L, Mastromei G (1999) Bacterial bio-mediated calcite precipitation for monumental stones conservation: methods of evaluation. J Microbiol Methods 36:139–145
Tiano P, Cantisani E, Sutherland I, Paget JM (2006) Bioremediated reinforcement of weathered calcareous stones. J Cult Herit 7:49e55
Troiano F, Vicini S, Gioventù E (2014) A methodology to select bacteria able to remove synthetic polymers. Polym Degrad Stab 107:321–327
Valentini F, Diamanti A, Palleschi G (2010) New bio-cleaning strategies on porous building materials affected by biodeterioration event. Appl Surf Sci 256:6550–6563
Valentini F, Diamantia A, Carbone M, Bauer EM, Palleschi G (2012) New cleaning strategies based on carbon nanomaterials applied to the deteriorated marble surfaces: a comparative study with enzyme based treatments. Appl Surf Sci 258(16):5965–5980
Vávrová P, Součková M (2017) Konzervace a RestaurováníNovodobýchKnihovníchFondů. Národníknihovna, Prague, p 197
Verstraete W, Wittebolle L, Heylen K, Vanparys B, de Vos P, van de Wiele T, Boon N (2007) Microbial resource management: the road to go for environmental biotechnology. Eng Life Sci 7:117–126
Vokic D, Berovic M (2005) Use of lipase to remove oil-based overpaints. ICOM- 13th Triennial meeting 12–16 September
Warscheid T, Braams J (2000) Biodeterioration of stone: a review. Int Biodeterior Biodegradation 46:343–368
Webster A, May E (2006) Bioremediation of weathered-building stone surfaces. Trends Biotechnol 24(6):255–260
Wendelbo O (1976) The use of proteolytic enzymes in the restoration of paper and papyrus. University Library of Bergen, Bergen
Wendelbo O, Fosse B (1970) Protein surgery: a restoring procedure applied on paper. Restaurator 1 (4):245–248
Wolbers R (2000) Cleaning painted surfaces: aqueous methods. Archetype Publications, London
Wolbers R (2007) Cleaning painted surface: aqueous methods. Archetype Books, London
Wolfaardt GM, Lawrence JR, Robarts RD, Caldwell DE (1998) In situ characterization of biofilm exopolymers involved in the accumulation of chlorinated organic. Microb Ecol 35:213–223
Zanardini E, Abbruscato P, Ghedini N, Realini M, Sorlini C (2000) Influence of atmospheric pollutants on the biodeterioration of stone. Int Biodeterior Biodegrad 45:35–42

Bioremediation of Cultural Heritage: Removal of Organic Substances

6

Chanda Parulekar-Berde, Rishikesh R. Surve, Sagar P. Salvi,
Prachiti P. Rawool, P. Veera Brahma Chari, and Vikrant B. Berde

Abstract

Cultural heritage (CH) deterioration is the combined effect of environmental factors and biological activities. Microorganisms grow on the art form establishing themselves and utilizing the available sources of carbon and energy. Consequences of these activities are seen in the form of damaged and deteriorating art forms, sometimes completely destroyed. Many methods are available for the cleaning and restoration of the damaged cultural heritage artwork. The use of microorganisms, i.e. biocleaning, amongst these, is the most effective and advantageous. The presence of organic matter in or on the CH, such as paintings, frescos and statues, renders them prone to microbial growth. The source of organic matter may be the activities of microorganisms growing on the art or organic matter used for restoration, for example, glue on paintings, or the composition of the art form itself such as parchment paper. Thus, the growth of microorganisms further degrades the surface of the CH. These artistic pathologies can be treated with help of viable microbial cultures. Selective isolation of non-pathogenic organic matter degraders is the starting point in biorestoration process. The potential of microorganisms for actual bioremediation of deteriorated cultural heritage materials is increasingly being unveiled and is the challenging field of research for aspirants.

C. Parulekar-Berde (✉) · S. P. Salvi · P. P. Rawool
Department of Microbiology, Gogate Jogalekar College, Ratnagiri, Maharashtra, India

R. R. Surve
Department of Chemistry, Arts, Commerce and Science College, Lanja, Maharashtra, India

P. V. B. Chari
Department of Biotechnology, Krishna University, Machilipatnam, Andhra Pradesh, India

V. B. Berde
Department of Zoology, Arts, Commerce and Science College, Lanja, Maharashtra, India

A. N. Yadav et al. (eds.), *Microbial Biotechnology Approaches to Monuments of Cultural Heritage*, https://doi.org/10.1007/978-981-15-3401-0_6

Keywords

Bioremediation · Cultural heritage · Monuments · Paintings · Biocleaning · Organic substances · Microorganisms

6.1 Introduction

The various types of artwork in cultural heritage (CH) include stonework, paintings and frescoes, which are prone to deterioration. This major concern with cultural heritage is the environmental factors involved in degradative actions as well as from living organisms causing deterioration. The artwork being exposed to environment, dust and air pollution over time gets damaged, and changes are seen in the form of deposition of dust, black crust formation, blackening, sulphation, etc. depending on the material of the art form, and their physical properties and susceptibility to deterioration differ accordingly (Saiz-Jimenez 1995; Webster and May 2006). In the case of monuments, for example, the materials used for construction may be marble, sandstone, quartz, etc. and therefore vary in porosity, alkalinity, hardness, etc.

Biodeterioration is the modification of organic and inorganic constituents brought about by the growth and metabolism of microorganisms. Biodeterioration plays a major role in the deterioration of artworks and monuments. The type of residual compounds added or present in the art form can act as growth substrates for microorganisms, under definite environmental conditions. Apart from this, several physical and chemical factors affect the biodeterioration process. Most significant abiotic factors include light intensity, air pollutants, humidity, temperature, solubility and porosity of the substratum, mechanical strength and nature of the material of the artwork and chemical nature of the material of CH. These factors can alter and destroy surfaces, resulting in changes in the colour and physical damage due to hyphal penetration (Ranalli et al. 1997; Tiano et al. 2006).

Though responsible for deterioration, microorganisms are looked upon as the new bioagents for the restoration and conservation of monuments and other artwork. The various anomalies in art form created due to microbial activities can be rectified using different cultures of viable bacteria.

Microorganisms, due to its specific enzyme production ability, have benefits over chemical methods and enzymes in the remediation of CH, in cleaning complex substances (nitrates, sulphates, organic substances) and material, which may be highly encrusted (Ranalli and Sorlini 2008). Thus, selecting an efficient microbial culture is the most important step in biorestoration of CH. Biocleaning technologies were applied to CH monuments like the Camposanto Monumentale cemetery in Pisa, Italy (Antonioli et al. 2005; Ranalli et al. 2005; Lustrato et al. 2012); the Santos Juanes Church in Valencia, Spain (Bosch-Roig et al. 2013a, b); the Duomo di Milano Cathedral in Milan, Italy (Cappitelli et al. 2006, 2007); the Matera Cathedral, Matera, Italy (May et al. 2008; Alfano et al. 2011); the Duomo di Firenze Cathedral in Florence, Italy (Gioventù et al. 2011); a nineteenth-century building in Riga,

Latvia; the Epidauro Theatre in Greece artworks like Michelangelo's 'Pietà Rondanini' and Lina Arpesani's sculpture, created in 1921 (Troiano et al. 2013); and original paper specimen from the Istituto Nazionale per la Grafica, Rome (removal of animal glue) (Barbabietola et al. 2012).

6.2 Damage to Cultural Heritage Due to Microorganisms

The environment can influence the colonization of microbes in a particular habitat. Regions having hot humid climate offer positive environmental conditions for microbial growth. These monuments get degraded and damaged following the growth of microorganisms that form a biofilm over the monument surface, thus eventually decreasing its aesthetic value (Gaylarde et al. 2012; Ortega-Calvo et al. 1991).

Various organic and inorganic components contained in artworks are responsible for creating an optimal environment for the colonization of microorganisms. Based on the type of constituents utilized by the organisms for metabolism, the nutritional groups were described (Caneva et al. 1991). Thus, autotrophic microorganisms that include the chemolithotrophs (sulphur-oxidizing and nitrifying bacteria) and photo-autotrophic bacteria (algae and cyanobacteria) are found on artwork like monuments, statues, etc., containing inorganic materials (Ranalli et al. 2009; Caneva et al. 1991). These are the primary colonizers. The secondary colonizers utilize the organic substances released by the lysis of primary colonizers or produced by them. The secondary colonizers include the heterotrophs, i.e. bacteria and fungi (Ranalli et al. 2009; Nuhoglu et al. 2006).

The bacterial genera *Flavobacterium*, *Pseudomonas*, *Alcaligenes*, *Bacillus* and *Staphylococcus* are commonly found to be associated with the deterioration of paintings (Capodicasa et al. 2010; Lopez-Miras et al. 2013; Pavic et al. 2015). Following the colonization of these bacterial species, fungal species start growing. Commonly found fungal species are *Aspergillus*, *Penicillium*, *Cladosporium* and *Alternaria* (Gorbushina et al. 2004; Caselli et al. 2018). The damage caused by fungal growth is often associated with deep mycelia penetration and paint discolouration, as a result of their metabolic activities. Thus, microbial communities colonizing the art forms and their interactive activities result in the deterioration of the art forms. The colonization and associative activities in the biofilm formed on art form, for example, are responsible for spoilage and deterioration in several CH materials (McNamara and Mitchell 2005). An example that showed the prevalence and involvement of microbial growth in deterioration was the study of the growth of autotrophic microbes on marble statue in the Boboli Gardens of Florence (Italy). Cyanobacteria *Coccomyxa* sp. was observed to colonize the surface, followed by a thin green biofilm of algae in a year of restoration (Lamenti et al. 2000).

6.2.1 Processes in the Deterioration of Cultural Heritage by Microorganisms

The numerous events occurring sequentially play a role in the biodeterioration of CH. These series of events involving abiotic and biotic factors result in the cumulative effect on the art forms.

Paintings: this art form includes three layers, i.e. the protective covering of varnish, then the picture and the support (Taft and Mayer 2000; Stulik 2000). The chemical composition of each layer differs depending upon the material utilized. This may include organic as well as inorganic components such as varnish, pigments, glue, oil, wood, paper, etc. (Stulik 2000; Leonardi 2005; Matteini and Mazzeo 2009).

Apart from being a source of nutrition to the microorganisms, the coating like varnish or glue also helps in the retention of water, necessary for microbial growth (Gaylarde and Gaylarde 2002; Giannantonia 2008). Salt efflorescence formation on the surface of wall paintings was reported. Perhaps, formation of salt effervescence is another reason leading to a deterioration of indoor CH. As the crystal increases in size, it exerts pressure on the walls of the paintings, resulting in traction forces and, eventually, the development of microcracks in the wall painting (Doménech-Carbó and Yusá-Marco 2006). Apart from salt swelling, filamentous fungi can also cause mechanical stress due to the penetration of the hyphae.

Stone monuments: outdoor CH are exposed to the detrimental effects of abiotic factors in the atmosphere such as gases. Nitrogen oxide and sulphur dioxide are predominant factors in causing damage and deterioration of exposed CH. Oxidation of these gases results in the formation of nitrous acid, nitric acid and sulphuric acid, which get deposited on to the surface of the monuments causing severe corrosion of the stone artwork. Sulphuric acid reacts with marble and other soluble calcareous substrates forming gypsum (Böke et al. 1999). The process of black crust formation takes place when during gypsum crystallization, organic pollutants and airborne particles get trapped on the surface of newly formed mineral matrix (Moropoulou et al. 1998; El-Metwally and Ramadan 2005). Organic substances trapped in the crusts can result in the growth of microorganisms, further degrading the CH (Kapsalas et al. 2007).

Nitrogen dioxide is oxidized to nitric acid, which reacts with the calcium carbonate present in the stone and forms calcium nitrate. Interestingly, this results in pulverized stone structures and microcrack formation in wall paintings (Warscheid and Braams 2000). Biotic factors are also involved in the deterioration process. Similar reactions occur due to the enzymes produced by *Thiobacillus* sp. and *Nitrobacter* sp. found colonizing the surfaces (Mitchell and Gu 2000; Nuhoglu et al. 2006). There are number of reports documenting the presence of nitrifying and sulphur-oxidizing bacteria on stone monuments (Spieck et al. 1992; BaumgaÈ rtner et al. 1991).

The biotic deterioration process starts with the initial colonizers. These grow on the surface of the art forms or may grow within. Phototrophic microorganisms including algae and cyanobacteria initiate colonization of microorganisms (Crispim et al. 2003; Negi and Sarethy 2019). Cyanobacteria and chlorophyta like *Gloeocapsa*, *Phormidium*, *Chorococcus*, *Chlorella*, *Stichococcus* and *Chlorococcum* constitute the pioneer species and primarily are responsible for the damage to cultural property (Macedo et al. 2009).

As the bacteria grow, it produces substances like exopolysaccharides that allow cell adhesion, entrapment of airborne particles and production of pigments, metal ions, minerals and organic compounds. Organic acids produced by the colonizers of the biofilm further result in salt formation due to reaction with the surface of stone or other structures. These salts ooze out and the process is called efflorescence (Scheerer et al. 2009), which ultimately results in crack formation. High porosity permits entry of microbes, ions and water inside the stone eventually causing severe damage (Dakal and Cameotra 2012; Saiz-Jimenez and Laiz 2000). Actinobacteria such as *Streptomyces* and *Nocardia* have been reported from the surface of stone murals (Ciferri 1999; Jurado et al. 2010).

6.3 Bioremediation of Cultural Heritage

Thus, microorganisms are known as causative agents of deterioration, and several reports are available indicating this fact (Saiz-Jimenez and Samson 1981; Montes and Hernández 1999; Cappitelli et al. 2004, Sofriti et al. 2019). However, the microorganisms do have positive effects also: they are used for biodeterioration and conservation of artwork (Saiz-Jimenez 1997; Webster and May 2006). Biocleaning using microorganisms has advantages over physico-chemical cleaning treatments and enzymes. Enzymes being substrate-specific are unable to degrade complex substances that are present on the deteriorating artworks. The physico-chemical treatments are drastic causing irreparable damage to the wall paintings. In-depth research based on the utility of microbiological systems for the removal of defects on artworks proves them to be alternatives for CH restoration.

The use of microorganisms for cleaning the surfaces of altered works of art has been applied to diverse weathering forms (Sorlini and Cappitelli 2008). The workers have reported the application of microbes for bioconsolidation and biocleaning of works of stone art. Biocleaning strategies involve both viable cells and enzymes to stop the deterioration of works of art and to restore the altered ones (Sorlini et al. 2010). In 2013, the cleaning of wall paintings in Santos Juanes Church in Valencia (Spain) was carried out using cells of *P. stutzeri*. This is the first report of biorestoration of nitrate salt efflorescence-affected wall painting (Bosch-Roig et al. 2013a).

Calcifying bacterium *Desulfovibrio desulfuricans* was used for the clearing of black crusts from stone. Atlas and Rude (1988) have described the effect of these calcifying bacteria on stone monuments. Due to the microbial activity, the accumulation of hydrogen carbonate and carbonate ions takes place on stone surface. However, this results indirectly in pH increase that favours $CaCO_3$ precipitation and stone consolidation (Castanier et al. 1999; Jroundi et al. 2012).

6.3.1 Criteria for Microorganisms to Be Used in Bioremediation Process

1. The microorganisms to be used for biorestoration should always be non-pathogenic and non-sporulating.
2. They should not present any health risks to either workers handling the microbial cultures, while in application, or the artworks.
3. The selected microorganisms should be able to adapt to different environments.
4. They should be able to produce enzymes as per the substrates present on the application site. This ensures its growth and also the removal of unwanted substances from the artwork (Ranalli et al. 2003).

6.3.2 Microorganisms Used for Bioremediation of Cultural Heritage

The application of the metabolic capabilities of the microbes has been defined as microbial resource management (MRM) (Verstraete et al. 2007). This new process of using bacterial cells in biorestoration is termed as biocleaning (Ranalli et al. 2000). To isolate effective biocleaning bacteria, it is recommended to use varnish, paint or polymer manufacturing industrial wastewater and sludge as source (Chen et al. 2007; Arutchelvan et al. 2005). Nonetheless, this ensures the selection of the appropriate microorganisms, having the ability to remove undesirable substances (Ranalli et al. 2000; Troiano et al. 2014).

Ranalli et al. (1996) were the first to apply nitrate-reducing bacteria (NRB) to bioclean Vicenza stone art altered by nitrates. The strain of *Pseudomonas stutzeri* was used for the trials wherein the bacterial culture was delivered in sepiolite.

Bellucci et al. 1999 reported the use of yeast *Candida cylindracea* lipase for removal of Paraloid B72 from two paintings. Similarly, there is another report of the application of a yeast culture found growing on a bronze statue treated with Incralac (copolymer of ethyl methacrylate and butyl acrylate), for degradation of the same coating (McNamara et al. 2004). Yeast being safe for the environment and user with low toxic metabolite production is suitable for biocleaning process (Saleem et al. 2008).

Sulphate reducers, *Desulfovibrio desulfuricans* and *Desulfovibrio vulgaris*, and nitrate-reducing bacteria, such as *Pseudomonas stutzeri*, *Pseudomonas denitrificans*, *Pseudomonas aeruginosa*, *Pseudomonas pseudoalcaligenes* and *Paracoccus denitrificans*, are promising biocleaning agents (Heselmeyer et al. 1991; Gauri et al. 1989, 1992; Ranalli et al. 1996, 2000, 2003, 2005; Cappitelli et al. 2005, 2006, 2007; Polo et al. 2010; Alfano et al. 2011).

P. stutzeri has only been applied for the removal of organic substances from wall paintings by a number of researchers (Ranalli et al. 2003, 2005; Antonioli et al. 2005; Sorlini and Cappitelli 2008; Polo et al. 2010).

6.3.3 Microorganism Application Methods and Delivery Systems Used

There are different ways of application of the microorganisms to the artwork, which include immersion in bacterial suspension, by direct bacterial application or by using delivery systems like cotton, cotton wool, agar, sepiolite, laponite, Carbogel and Hydrobiogel-97 (Ranalli et al. 1997; Antonioli et al. 2005; Cappitelli et al. 2006; Alfano et al. 2011). The selection of delivery system has to follow certain guidelines to ensure successful cleaning of organic substances from the artwork. The most important feature of biocleaning is the delivery system followed. The criteria for selection of delivery system are as follows (Bosch-Roig et al. 2014; Bosch-Roig and Ranalli 2014):

1. It should provide the bacteria with the right environmental conditions, for example, supply of enough water (Campani et al. 2007).
2. It should keep the bacteria in contact with the degraded or affected part of the artwork, without interacting with the surface.
3. It should be easily applicable to all types of surfaces
4. It should be easy and quick to prepare, apply and remove.
5. It should be prepared using only a few inexpensive materials.

Thus, systems retaining less water and long applications were required for strongly degraded and sensitive surfaces (such as wall paintings). An example is the use of agar to deliver bacteria on Japanese paper. In case of strong application and need of homogeneity, for example, for applying anaerobic bacteria, Carbogel was used. During the first trials with a strain of *Pseudomonas pseudoalcaligenes* for biocleaning, Carbogel or a mortar-alginate matrix was used as a delivery system (70, 75). The use of Carbogel was reported for specific strains of *Desulfovibrio* spp. and *Desulfovibrio vulgaris* subsp. *vulgaris* (ATCC 29579) (Cappitelli et al. 2005, 2006, 2007; Giovent and Lorenzi 2013).

For biorestoration of vertical surfaces, micro-packs of laponite are employed. The living bacterial cells were mixed in a laponite gel and applied onto the surface of the artwork. It possesses all the properties of a good carrier (Mazzoni et al. 2014). For artwork with pre-consolidated surfaces, cost-effective applications may be obtained using cotton wool, which do not pose risk to the artworks and easy-to-use applications (Ranalli et al. 2005; Lustrato et al. 2012; Sanmartín and Bosch-Roig 2019). There are reports of the biocleaning of nitrate salt efflorescence from wall paintings by using *P. stutzeri*, using agar as the delivery system (Ranalli et al. 2019; Bosch-Roig et al. 2013a, b; Sanmartín and Bosch-Roig 2019).

6.4 Microbial Removal of Organic Substances Damaging Cultural Heritage

Accumulation of organic matter in/on CH artwork results in the growth of microorganisms leading to the deterioration of these art forms. Thus, removal of the organic substances from the art forms is a major step in the restoration of CH. The various organic substances found are casein, egg yolk, oil and animal fat, which are difficult to remove with other techniques. The technique requires shorter application time and, along with effectiveness, should be easy to apply (Ranalli et al. 2019). Indeed the modern cleaning techniques such as the enzymes, surfactants and solubilizing agents are a little less efficient when it comes to restoration of frescoes (Makes 1988; Bellucci and Cremonesi 1994; Bonomi 1994; Wolbers 2007). However, Ranalli and colleagues have shown the failure of these methods based on pioneering research on the Pisa frescoes (Ranalli et al. 2005, 2009).

The application of whole cells of microorganisms is thus the best methodology. Report of Ranalli et al. (2009) on the animal glue bioremoval from a Spinello Aretino fresco in the Monumental Cemetery of the Camposanto in Pisa using *Pseudomonas stutzeri* is an example of positive results obtained using microorganisms.

6.4.1 Methods of Biocleaning Organic Substances

The ability of microorganisms for effective bioremediation of deteriorated CH materials is gradually unveiled, and promising results of this methodology are reported from field sites. The use of microorganisms to deal with removal of organic matter from stone surfaces during restoration interventions has been reported.

The biocleaning of organic substances from the wall paintings, using the bacterial culture *P. stutzeri*, has only been reported (Ranalli et al. 2003, 2005; Antonioli et al. 2005; Sorlini and Cappitelli 2008; Polo et al. 2010). These reports were based on detailed studies involving the application of bacteria to the art forms and the methodologies utilized. The CH was exposed to bacteria in different ways. One method was the direct immersion of the art form in culture suspension; another method described was a direct application of bacteria, and the most effective method is the application of bacterial culture using the delivery system. The different delivery systems used are cotton, sepiolite, Hydrobiogel-97, Carbogel and multilayer systems (Ranalli et al. 1997; Antonioli et al. 2005; Cappitelli et al. 2006; Alfano et al. 2011).

The methodology used for the removal of proteinaceous residues, such as glue and casein, the remains of the old restoration process, is described below:

1. Organic matter attached to the artwork analysed by physico-chemical methods
2. Development of biocleaning method for the removal of organic substances

3. Monitoring the biocleaning agents, i.e. the microorganisms, during the biocleaning period

The actual biocleaning method requires enough biomass, a good monitoring method and complete removal of the organic matter. After the biocleaning process, absence or removal of biocleaning agent should be ensured.

6.4.2 Examples of Biocleaning Organic Substances

Best known types of artwork, have been affected by microbial and physico-chemical agents of deterioration. Methods of restoration eventually may also be reasons for deterioration such as the application of organic matter used for biorestoration (Ranalli et al. 1997; Tiano et al. 2006).

Organic substances are causative agents of deterioration as they facilitate the growth of microorganisms. Microorganisms also grow on artwork that supports growth and may produce organic substances that lead to spoilage. The use of beneficial microorganisms for the biocleaning and biorestoration processes, especially the removal of organic substances, has proven to be the best and safe options so far.

Mazzoni et al. (2014) reported the use of microorganisms for the removal of calcium carbonate, gypsum, weddellite, apatite and casein deposits, from the wall paintings of the lower loggia of the Casina Farnese (Palatine Hill, Rome). *Stenotrophomonas maltophilia* was applied for removal of organic deposits, while gypsum, carbonates and phosphates were solubilized by *Cellulosimicrobium cellulans* TBF11^{E} and *Pseudomonas koreensis* UT30^{E}. The delivery system used for applying the organisms was laponite, in the form of *micro*-packs. The biocleaning of organic matter on artwork for the removal of casein and glues on frescoes using living microorganisms has been previously described (Lustrato et al. 2012; Ranalli et al. 2005).

In addition to bacteria, the use of yeast in biorestoration has also been reported. *Candida cylindracea* producing a lipase enzyme has been used for the removal of Paraloid B72 from paintings (Bellucci et al. 1999). Another report is about the isolation of a yeast culture from an Incralac (an ethyl methacrylate and butyl acrylate copolymer)-treated bronze statue. This culture helped in the removal of the coating itself that was the cause of deterioration of the art form (McNamara et al. 2004).

Ranalli et al. (2005) carried out the biorestoration work on the fourteenth-century fresco painting 'Stories of the Holy Fathers' by Buffalmacco Buonamico at Camposanto Monumentale in Pisa, Italy. Notably, this was the innovative work wherein viable cells of *P. stutzeri* A29 strain were used to remove altered animal glue. This application was of record time of 2 hours, the shortest reported (Lustrato et al. 2012). *Pseudomonas stutzeri* was successfully used for the biocleaning of organic matter from paintings damaged during the Second World War (Ranalli et al. 2003, 2005; Antonioli et al. 2005; Sorlini and Cappitelli 2008; Polo et al. 2010). Bacteria are excellent biorestoration agents on the fresco (Ranalli et al. 2018).

6.5 Advantages of Biocleaning Methods

Numerous physical and chemical methods were prescribed for organic matter removal from artwork. These methods are aggressive, invasive and non-selective, as well as time-consuming. The chemicals used in the process may be toxic to the workers involved, thus risky for humans as well as the environment. Application of these methods leads to colour variations, salt movements in the material structure and alterations in the art form due to the removal of material of art form, along with unwanted material (Cappitelli et al. 2007). Hence, these methods have a number of drawbacks along with being ineffective in most cases, in comparison with the biocleaning methods using microorganisms.

Biocleaning offers many advantages over the previously used methods of restoration: (Vergès-Belmin 1996; Webster and May 2006)

- Selectivity – depending on the type of organic substance to be removed, the microorganism may be used selectively. The organism will not act on the art form or another material present.
- The use of safe non-pathogenic microorganisms ensures the safety of conservators, operators as well as the environment.
- Biocleaning methods are non-toxic, non-aggressive and non-invasive, as well as highly specific, and cause no pollution as there is no release of any toxic chemicals in the environment.
- Relatively low cost as compared to physical, chemical or enzymatic methods. The use of enzymes, such as protease and collagenase, is more expensive than using the whole cells (Ranalli et al. 2005).
- The use of viable microbial culture is more efficient as it provides high homogeneity in the removal of deposits of organic matter, without harming the art forms (Cappitelli et al. 2007; Sorlini and Cappitelli 2008).
- Biocleaning is non-destructive and cleans the artwork of only compounds produced by spoilage organisms and remains of substances used during previous renovation work. The artwork remains unharmed.

Biocleaning being a sustainable approach is a way to go green in CH conservation practice (Cappitelli 2016).

6.6 Conclusion and Future Prospects

The biocleaning methods are however still not used routinely in CH restoration. Though it seems to be most effective, there are a number of issues that need to be addressed. The methods vary depending on the material of artwork or the type of matter to be removed, the application methods need to be more developed or standardized and so on. The potentiality of bioremediation processes, using microorganisms, for the restoration of artwork especially in the removal of organic matter, has been proven. However, it also gives rise to the question of risks of exposure of artwork, workers and environment to the microorganisms in use. Thus, reassuring research to remove this reluctance is the need of the day. Recently studies have proven promising for a selective and environmental/health-safe approach in the cultural heritage field (Lustrato et al. 2012; Ranalli et al. 2000; Bellucci et al. 1999). Consequently, every study that relates to the use of viable microorganisms for biorestoration and bioconservation is an advancement in the field of conservation and protection of CH. Based on studies pursued in Pisa, the application of *Pseudomonas stutzeri* in pure cultures has been accepted by the Technical Commission for Restoration (Pisa), and the bioremediation process is still being extensively used in restoration processes (Ranalli et al. 2000).

Acknowledgement The authors are thankful to their respective institutions for encouragement and support.

References

Alfano G, Lustrato G, Belli C, Zanardini E, Cappitelli F, Mello E, Sorlini C, Ranalli G (2011) Biodegradation, the bioremoval of nitrate and sulfate alterations on artistic stonework: the case-study of Matera Cathedral after six years from the treatment. Int Biodeterior Biodegrad 65:1004–1011

Antonioli P, Zapparoli G, Abbruscato P, Sorlini C, Ranalli G, Righetti PG (2005) Art-loving bugs: the resurrection of Spinello Aretino from Pisa's cemetery. Proteomics 5:2453–2459

Arutchelvan V, Kanakasabai V, Nagarajan S, Muralikrishnan V (2005) Isolation and identification of novel high strength phenol degrading bacterial strains from phenol-formaldehyde resin manufacturing industrial wastewater. J Hazard Mater 127:238–243

Atlas RC, Rude PO (1988) Complete oxidation of solid phase sulfides by manganese and bacteria in anoxic marine sediment. Geochim Cosmochim Acta 52:751–766

Barbabietola N, Tasso F, Grimald M, Alisi C, Chiavarini S, Marconi P et al (2012) Microbe-based technology for a novel approach to conservation and restoration. EAI Speciale II. Knowl Diagn Preserv Cult Herit 2012:69e76

BaumgaÈ rtner M, Sameluck F, Bock E, Conrad R (1991) Production of nitric oxide by ammonium-oxidizing bacteria colonizing building stones. FEMS Microbiol Ecol 8:95–100

Bellucci R, Cremonesi P (1994) L'uso degli enzimi nella conservazione e nel restauro dei dipinti. Kermes 21:45–64

Bellucci R, Cremonesi P, Pignagnoli G (1999) A preliminary note on the use of enzymes in conservation: the removal of aged acrylic resin coatings with lipase. Stud Conserv 44:278–281

Böke H, Gokturk EH, Caner-Saltik EN, Demirci S (1999) Effect of airborne particle on SO2-calcite reaction. Appl Surf Sci 140:70–82

Bonomi R (1994) Utilizzo degli enzimi per il restauro di una scultura in terracotta policroma. OPD Restauro 6:101–107

Bosch-Roig P, Ranalli G (2014) The safety of biocleaning technologies for cultural heritage. Front Microbiol 5:155

Bosch-Roig P, Regidor-Ros JL, Montes-Estellés R (2013a) Biocleaning of nitrate alterations on wall paintings by *Pseudomonas stutzeri*. Int Biodeterior Biodegrad 84:266–274

Bosch-Roig P, Regidor-Ros JL, Soriano-Sancho P, Montes-Estelles RM (2013b) Biocleaning of animal glue of wall paintings by *Pseudomonas stutzeri*. Chim Oggi-Chem Today 31:50–53

Bosch-Roig P, Lustrato G, Zanardini E, Ranalli G (2014) Biocleaning of cultural heritage stone surfaces and frescoes: which delivery system can be the most appropriate? Ann Microbiol 65:1227–1241

Campani E, Casoli A, Cremoneso C, Saccani I, Signorini E (2007) In: Prato Il (ed) L'uso di agarosio e agar per la preparazione di "Gel rigidi", vol 4. Cesmar, Padova

Caneva G, Nugari MP, Salvadori O (1991) Environmental factors in biodeterioration. In: Caneva G, Nugari MP, Salvadori O (eds) Biology in the conservation of works of art. ICCROM, Rome, pp 3–24

Capodicasa S, Fedi S, Porcelli AM, Zannoni D (2010) The microbial community dwelling on a biodeteriorated 16th century painting. Int. Biodeterior Biodegrad 64:727–733

Cappitelli F (2016) Biocleaning of cultural heritage surfaces. Open Conf Proc J 7:65–69

Cappitelli F, Zanardini E, Sorlini C (2004) The biodeterioration of synthetic resins used in conservation. Macromol Biosci 4:399–406

Cappitelli F, Zanardini E, Toniolo L, Abbruscato P, Ranalli G, Sorlini C (2005) Bioconservation of the marble base of the Pietà Rondanini by Michelangelo Buonarroti. J Appl Microbiol 7:06675

Cappitelli F, Principi P, Sorlini C (2006) Biodeterioration of modern materials in contemporary collections: can biotechnology help? Trends Biotechnol 24:350–354

Cappitelli F, Toniolo L, Sansonetti A, Gulotta D, Ranalli G, Zanardini E, Sorlini C (2007) Advantages of using microbial technology over traditional chemical technology in removal of black crusts from stone surfaces of historical monuments. Appl Environ Microbiol 73:5671–5675

Caselli E, Pancaldi S, Baldisserotto C, Petrucci F, Impallaria A, Volpe L, D'Accolti M, Soritti I, Coccagna M, Sassu G et al (2018) Characterization of biodegradation in a 17th century easel painting and potential for a biological approach. PLoS One 13:e0207630

Castanier S, Le Metayer-Levrel G, Perthuisot JP (1999) Ca-carbonates precipitation and limestone genesis-the microbiogeologist point of view. Sediment Geol 126:9–23

Chen J, Zhang Y, Du GC, Hua ZZ, Zhu Y (2007) Biodegradation of polyvinyl alcohol by a mixed microbial culture. Enzym Microb Technol 40:1686–1691

Ciferri O (1999) Microbial degradation of paintings. Appl Environ Microbiol 65:879–885

Crispim C, Gaylarde P, Gaylarde C (2003) Algal and cyanobacterial biofilms on calcareous historic buildings. Curr Microbiol 46:79–82

Dakal TC, Cameotra SS (2012) Microbially induced deterioration of architectural heritages: routes and mechanisms involved. Environ Sci Eur 24:36

Doménech-Carbó MT, Yusá-Marco DJ (2006) Aspectos físicoquímicos de la pintura mural y su limpieza. Polytechnic University of Valencia, Valencia

El-Metwally AA, Ramadan AB (2005) The role of air pollutants and sewage waste in acceleration of degradation of the Islamic cultural heritage of Cairo. In: Comparative risk assessment and environmental decision making, vol 38, part 3. Springer, Dordrecht, pp 363–370

Gauri LK, Chowdhury AN, Kulshreshtha NP, Punuru AR (1989) The sulfation of marble and the treatment of gypsum crusts. Stud Conserv 34:201–206

Gauri LK, Parks L, Jaynes J, Atlas R (1992) Removal of sulphated crust from marble using sulphate-reducing bacteria. In: Robin GM (ed) Stone cleaning and the nature, soiling and decay mechanisms of stone. Proceedings of the International Conference, 14 to 16 April 1992. Donhead Publishing Ltd, Webster, Edinburgh, pp 160–165

Gaylarde C, Gaylarde P (2002) Biodeterioration of historic buildings in Latin America. DBMC 9:171–180

Gaylarde CC, Rodrıguez CH, Navarro NY, Ortega MB (2012) Microbial biofilms on the sandstone monuments of the Angkor Wat Complex, Cambodia. Curr Microbiol 64:85–92

Giannantonia D (2008) Molecular characterization of microbial communities fouling concrete infrastructures. Dissertation, Georgia Institute of Technology

Gioventù E, Lorenzi P (2013) Bio-removal of black Crust from marble surface: comparison with traditional methodologies and application on a sculpture from the Florence's English Cemetery. Proc Chem 8:123–129

Gioventù E, Lorenzi PF, Villa F, Sorlini C, Rizzi M, Cagnini A et al (2011) Comparing the bioremoval of black crusts on colored artistic lithotypes of the cathedral of Florence with chemical and laser treatment. Int Biodeterior Biodegradation 65:832–839

Gorbushina AA, Heyrman J, Dornieden T, Gonzalez-Delvalle M, Krumbein WE, Laiz L, Petersen K, Saiz-Jimenez C, Swings J (2004) Bacterial and fungal diversity and biodeterioration problems in mural painting environments of St. Martins church (Greene–Kreiensen, Germany). Int Biodeterior Biodegrad 53:13–24

Heselmeyer K, Fischer U, Krumbein WE, Warsheid T (1991) Application of *Desulfovibrio vulgaris* for the bioconversion of rock gypsum crusts into calcite. BIOforum 1:89

Jroundi F, Gomez-Suaga P, Jimenez-Lopez C, Gonzalez-Munoz MT, Fernandez-Vivas MA (2012) Stone isolated carbonatogenic bacteria as inoculants in bioconsolidation treatments for historical limestone. Sci Total Environ 425:89–98

Jurado V, Laiz L, Rodriguez-Nava V, Boiron P, Hermosin B, Sanchez-Moral S et al (2010) Pathogenic and opportunistic microorganisms in caves. Int J Speleol 39:15–24

Kapsalas P, Zervakis M, Maravelaki-Kalaitzaki P (2007) Evaluation of image segmentation approaches for non-destructive detection and quantification of corrosion damage on stonework. Corros Sci 49:4415–4442

Lamenti G, Tiano P, Tomaselli L (2000) Biodeterioration of ornamental marble statues in the Boboli gardens (Florence, Italy). J Appl Phycol 12:427–433

Leonardi R (2005) Nuclear physics and painting: sub-topic of the wide and fascinating field of science and art. Nucl Phys A 752:659–674

Lopez-Miras MD, Martin-Sanchez I, Yebra-Rodriguez A, Romero-Noguera J, Bolivar-Galiano F, Ettenauer J, Sterflinger K, Pinar G (2013) Contribution of the microbial communities detected on an oil painting on canvas to its biodeterioration. PLoS One 8:e80198

Lustrato G, Alfano G, Andreotti A, Colombini MP, Ranalli G (2012) Fast biocleaning of mediaeval frescoes using viable bacterial cells. Int. Biodeterior Biodegrad 69:51–61

Macedo M, Miller A, Dionisio A, Saiz-Jimenez C (2009) Biodiversity of cyanobacteria and green algae on monuments in the Mediterranean Basin: an overview. Microbiology 155:3476–3490

Makes F (1988) Enzymatic consolidation of the portrait of Rudolf II as _Vertumnus_ by Giuseppe Arcimboldo with a new multi-enzyme preparation isolated from Antarctic Krill (Euphasia superba). Acta Universitatis Gotheburg 23:98–110

Matteini M, Mazzeo R (2009) Structure of panel and canvas paintings. In: Pinna D, Galeotti M, Mazzeo R (eds) Scientific examination for the investigation of paintings, a handbook for conservator-restorers. Centro Di, Firenze, pp 11–20

May E, Webster AM, Inkpen R, Zamarreno D, Kuever J, Rudolph C, Warscheid T, Sorlini C, Cappitelli F, Zanardini E et al (2008) The biobrush project for bioremediation of heritage stone. In: May E, Jones M, Mitchell J (eds) Heritage microbiology and science: microbes, monuments and maritime materials. RSC Publishing, Cambridge, pp 76–93

Mazzoni M, Alisi C, Tasso F, Cecchini A, Marconi P, Sprocati AR (2014) Laponite micro-packs for the selective cleaning of multiple coherent deposits on wall paintings: the case study of Casina Farnese on the Palatine Hill (Rome-Italy). Int Biodeterior Biodegrad 94:1–11

McNamara CJ, Mitchell R (2005) Microbial deterioration of historic stone. Front Ecol Environ 3:445–451

McNamara CJ, Breuker M, Helms M, Perry TD, Mitchell R (2004) Biodeterioration of Incralac used for the protection of bronze monuments. J Cult Herit 5:361–364

Mitchell R, Gu JD (2000) Changes in the biofilm microflora of limestone caused by atmospheric pollutants. Int Biodeterior Biodegradation 46:299–303

Montes R, Hernández E (1999) Estudio de la contaminación microbiológica en la pinturas murales de la Basílica de la Virgen de los Desamparados de Valencia. In: Roig P, Bosch I (eds) Restauración de Pintura Mural aplicada a la Basílica de al Virgen de los Desamparados de Valencia. Universidad Politécnica de Valencia, Valencia, pp 201–215

Moropoulou A, Bisbikou K, Torfs K, Van Grieken R, Zezza F, Macri F (1998) Origin and growth of weathering crusts on ancient marbles in industrial atmosphere. Atmos Environ 32:967–982

Negi A, Sarethy IP (2019) Microbial biodeterioration of cultural heritage: events, colonization, and analyses. Microb Ecol 78:1014

Nuhoglu Y, Oguz E, Uslu H, Ozbek A, Ipekoglu B, Ocak I et al (2006) The accelerating effects of the microorganisms on biodeterioration of stone monuments under air pollution and continental-cold climatic conditions in Erzurum, Turkey. Sci Total Environ 364:272–283

Ortega-Calvo JJ, Naturales R, Saiz-Jiminez C (1991) Biodeterioration of building materials by cyanobacteria and algae. Int Biodeterior 28:165–185

Pavic A, Ilic-Tomic T, Pacevski A, Nedeljkovic T, Vasiljevic B, Moric I (2015) Diversity and biodeteriorative potential of bacterial isolates from deteriorated modern combined-technique canvas painting. Int Biodeterior Biodegrad 97:40–50

Polo A, Cappitelli F, Brusetti L, Principi P, Villa F, Giacomucci L, Ranalli G, Sorlini C (2010) Feasibility of removing surface deposits on stone using biological and chemical remediation methods. Microb Ecol 60:1–14

Ranalli G, Sorlini C (2008) Bioremediation. In: Caneva G, Nugari MP, Salvadori O (eds) Plant biology for cultural heritage. Bioremediation and conservation. The Getty Conservation Institute, Los Angeles, pp 340–346

Ranalli G, Chiavarini M, Guidetti V, Marsala F, Matteini M, Zanardini E, Sorlini C (1996) The use of microorganisms for the removal of nitrates and organic substances on artistic stoneworks. In: Riederer J (ed) Proceedings 8th international congress on deterioration and conservation of stone. Mo¨ller, Berlin, pp 1415–1420

Ranalli G, Chiavarini M, Guidetti V, Marsala F, Matteini M, Zanardini E, Sorlini C (1997) The use of microorganisms for the removal of sulphates on artistic stoneworks. Int Biodeterior Biodegrad 40:255–261

Ranalli G, Matteini M, Tosini I, Zanardini E, Sorlini C (2000) Bioremediation of cultural heritage: removal of sulfates, nitrates and organic substances. In: Ciferri O, Tiano P, Mastromei G (eds) Of microbes and art: bioremediation of cultural heritage: removal of sulphates, nitrates and organic substances. Springer, New York

Ranalli G, Belli C, Baracchini C, Caponi G, Pacini P, Zanardini E, Sorlini C (2003) Deterioration and bioremediation of frescoes: a case-study. In: Saiz-Jimenez C (ed) Molecular biology and cultural heritage. A.A. Balkema Publishers, Lisse, pp 243–246

Ranalli G, Alfano G, Belli C, Lustrato G, Colombini MP, Bonaduce I, Zanardini E, Abbruscato P, Cappitelli F, Sorlini C (2005) Biotechnology applied to cultural heritage: biorestoration of frescoes using viable bacterial cells and enzymes. J Appl Microbiol 98:73–83

Ranalli G, Zanardini E, Sorlini C (2009) Biodeterioration—including cultural heritage. In: Schaechter M (ed) Encyclopedia of microbiology. Academic Press, Oxford, pp 191–205

Ranalli G, Zanardini E, Andreotti A, Colombini MP, Corti C, Bosch-Roig P, De Nuntiis P, Lustrato G, Mandrioli P, Rampazzi L et al (2018) Hi-tech restoration by two-steps biocleaning process of Triumph of Death fresco at the Camposanto Monumental Cemetery (Pisa, Italy). J Appl Microbiol 125:800–812

Ranalli G, Zanardini E, Rampazzi L, Corti C, Andreotti A, Colombini MP, Bosch-Roig P, Lustrato G, Giantomassi G, Zari D et al (2019) Onsite advanced biocleaning system on historical wall paintings using new agar-gauze bacteria gel. J Appl Microbiol 126:1785–1796

Saiz-Jimenez C (1995) Deposition of anthropogenic compounds on monuments and their effect on airborne microorganisms. Aerobiologia 11:161–175

Saiz-Jimenez C (1997) Biodeterioration vs. biodegradation: the role of microorganisms in the removal of pollutants deposited on historic buildings. Int Biodeterior Biodegrad 24:225–232

Saiz-Jimenez C, Laiz L (2000) Occurrence of halotolerant/halophilic bacterial communities in deteriorated monuments. Int Biodeterior Biodegrad 46:319–326

Saiz-Jimenez C, Samson RA (1981) Biodegradación de obras de arte. Hongos implicados en la degradación de los frescos del monasterio de la Rábida (Huelva). Botanica Macaronesica 8–9:255–264

Saleem M, Brim H, Hussain S, Arshad M, Leigh MB (2008) Perspectives on microbial cell surface display in bioremediation. Biotechnol Adv 26:151–161

Sanmartín P, Bosch-Roig P (2019) Biocleaning to remove graffiti: a real possibility? advances towards a complete protocol of action. Coatings 9:104

Scheerer S, Ortega-Morales O, Gaylarde C (2009) Microbial deterioration of stone monuments—an updated overview. Adv Appl Microbiol 66:97–139

Sofritti I, D'Accolti M, Lanzoni L, Volta A, Bisi M, Mazzacane S, Caselli E (2019) The potential use of microorganisms as restorative agents: an update. Sustainability 11:3853–3870

Sorlini C, Cappitelli F (2008) The application of viable bacteria for the biocleaning of cultural heritage surfaces. Coalition 15:18e20

Sorlini C, Ranalli G, Zanardini E (2010) Stone bioconservation. In: Mitchell R, McNamara CJ (eds) Cultural heritage microbiology: fundamental studies in conservation science. ASM Press, Washington, DC, pp 291–303

Spieck E, Meincke M, Bock E (1992) Taxonomic diversity of *Nitrosovibrio* strains isolated from building sandstones. FEMS Microbiol Ecol 102:21–26

Stulik D (2000) Paint. In: WSJ T, Mayer JW (eds) The science of paintings. Springer, New York, pp 12–25

Taft WSJ, Mayer JW (2000) The structure and analysis of paintings. In: WSJ T, Mayer JW (eds) The science of paintings. Springer, New York, pp 1–11

Tiano P, Cantisani E, Sutherland I, Paget JM (2006) Bio remediated reinforcement of weathered calcareous stones. J Cult Herit 7:49e55

Troiano F, Guiotta D, Balloi A, Polo A, Toniolo L, Lombardi E, Daffonchio D, Sorlini C, Cappitelli F (2013) Successful combination of chemical and biological treatments for the cleaning of stone artworks. Int Biodeterior Biodegrad 85:294–304

Troiano F, Vicini S, Gioventù E (2014) A methodology to select bacteria able to remove synthetic polymers. Polym Degrad Stab 107:321–327

Vergès-Belmin V (1996) Towards a definition of common evaluation criteria for the cleaning of porous building materials: a review. Sci Technol Cult Herit 5:69–83

Verstraete W, Wittebolle L, Heylen K, Vanparys B, de Vos P, van de Wiele T, Boon N (2007) Microbial resource management: the road to go for environmental biotechnology. Eng Life Sci 7:117–126

Warscheid T, Braams J (2000) Biodeterioration of stone: a review. Int Biodeterior Biodegradation 46:343–368

Webster A, May E (2006) Bioremediation of weathered-building stone surfaces. Trends Biotechnol 24:255–260

Wolbers R (2007) Cleaning painted surface: aqueous methods. Archetype Books, London

The Role of Microorganisms in Removal of Sulfates from Artistic Stonework

7

Prem Chandra, Enespa, Rajesh Kumar, and Jameel Ahmad

Abstract

The presence of organic matter on artistic stonework can be credited to inadequate historical renovations, lysis of microbial cells, primary surface colonization, and manifestation of hydrocarbons from oil combustion. The conservation of the artwork itself can be seriously dangerous. To date, surfactants and solubilizing agents have been used to remove pollutants and residual substances from artwork by chemical and physical procedures. For biological removal of sulfates, nitrates, and organic matter present on artistic stonework, multiple bioremediation systems have now been developed, exploiting prudently selected microbial cultures grown on an appropriate support. The development of this process involves screening and selection of a suitable microbial culture with the capability to biodegrade organic matter, denitrify, and reduce sulfates; setting up of simulated laboratory tests with stone samples artificially enriched with nitrates, sulfates, and organic matter; testing of appropriate inert matrices on which to immobilize the selected bacterial strains; and testing of sulfate, nitrate, and organic matter removal from artificially enriched stone, as well as from naturally degraded artwork. Bacterial biofilms using sepiolite with a high active biomass per cm^3 were developed. *Pseudomonas aeruginosa* and *Pseudomonas stutzeri* were selected for nitrate removal because of their high denitrifying activity. *Desulfovibrio vulgaris* and *Desulfovibrio desulfuricans* were selected and tested

P. Chandra (✉)
Department of Environmental Microbiology, Babasaheb Bhimrao Ambedkar University (A Central University), Lucknow, Uttar Pradesh, India

Enespa
Department of Plant Pathology, School of Agriculture, Sri Mahesh Prasad Degree College, University of Lucknow, Lucknow, India

R. Kumar · J. Ahmad
Department of Zoology, Gandhi Faizam College, Mahatama Jyotiba Phule Rohilkhand University, Bareilly, Uttar Pradesh, India

A. N. Yadav et al. (eds.), *Microbial Biotechnology Approaches to Monuments of Cultural Heritage*, https://doi.org/10.1007/978-981-15-3401-0_7

for sulfate removal in liquid cultures, on stone specimens artificially enriched with sulfates, and on real marble samples. The results confirmed the potential for development of bioremediation as a soft, innovative technology based on the use of microorganisms and their metabolic activity in recovery of degraded artworks.

Keywords

Bioremediation of artworks · *Desulfovibrio vulgaris* · *Desulfovibrio desulfuricans* · Nitrates · Sulfates · Organic matter

7.1 Introduction

Sulfate components are generated from the oxidation of sulfur dioxide into sulfur trioxide and finally form sulfuric acid. (Constantin et al. 1996; Larson et al. 2000; Ranalli et al. 2000). The fraction of the sulfur component that is not neutralized in air contributes to the materialization of acid rain (Wang et al. 2016; Velikova et al. 2000). The steady increase in rainwater acidity over the past few years poses a prospective threat to humans, nature, and sculptures by instigating the alteration of intricate calcium carbonate into more soluble calcium sulfate dehydrate or gypsum ($CaSO_4{\cdot}2H_2O$) (Urzì 1999; Van Driessche et al. 2017). Carbonaceous particles are intertwined and form black spots in sheltered areas in a mineral medium after materialization of crystallized gypsum from floating pollutants (such as spores, pollen, dust particles, and different particulate materials—known as smog—including heavy hydrocarbons such as those in the aromatic and aliphatic series). Mostly, it is assumed that they are engendered by wet and dry deposition procedures in which sulfuric acid, an oxidation product of sulfur dioxide, attacks carbonic rocks and the procedure of gypsum formation results (Buckman and Brady 1960; Brady and Weil 1999):

$$2SO_2 + O_2 \rightarrow 2SO_3 \quad (7.1)$$

$$SO_3 + H_2O \rightarrow H_2SO_4 \quad (7.2)$$

$$CaCO_3 + H_2SO_4 \rightarrow CaSO_4 \cdot 2H_2O + CO_2 \quad (7.3)$$

None of the available mechanical and chemical treatments invented for the cleaning of stone altered by black crusts have proved to be entirely acceptable (Cappitelli et al. 2006). Conclusively, the presence of organic matter on creative stonework can be attributable to inadequate former refurbishment, lysis of microbial cells from primary surface colonization, and the presence of hydrocarbons originating from oil combustion (Pereira Roders and Van Oers 2011), representing serious threats to preservation of artworks. Artistic stonework is sited in the open atmosphere, where the influence of atmospheric effluence speeds up the degradation of materials (Fernandes 2006; Kuhad and Singh 2013). Among the environmental factors that cause such processes are potential hydrogen (pH), humidity, wind, rainfall, thermal variations, and organic and inorganic pollutants (Syed 2006). The pathologies of materials constituting artworks are generally alleviated by chemical or physical techniques; biological methods are rarely used (Kety 1979). Actually, only some enzymatic activities have been applicable to the retrieval of some reforms. Whole living microbes as agents of biorecovery have never been utilized; only sporadic studies under laboratory conditions have been carried out (Ciferri et al. 2000; Palková 2004).

In the aforementioned studies, experiments were carried out using sulfate-reducing bacteria in broth cultures on altered stone samples (Parkes et al. 1989; Martins et al. 2009). This process uses anaerobic bacteria capable of reducing sulfate to hydrogen sulfide, which, in turn, is emitted into the atmosphere (Ayangbenro et al. 2018). The bacterial genera of *Desulfovibrio*, *Desulfomonas*, *Desulfobacter*, *Desulfococcus*, and *Desulfosarcina* are naturally present in anaerobic mud and freshwater residue, in marine environments, and in the gastrointestinal tracts of humans and animals (Rabus et al. 2013; Kim et al. 1999). *Desulfovibrio desulfuricans*, a sulfate-reducing bacterium, was used in aliquots of broth to treat marble samples showing black weathering crusts rich in gypsum (Atlas et al. 1988; Ciferri et al. 2000). Calcite was found on all treated surfaces described in various reports recommending prospective use of this microorganism to clean coated marble (Heggendorn et al. 2018; Md Zain et al. 2018). When bacteria reduce sulfates, calcium ions (Ca^{2+}) released from gypsum react with carbon dioxide, resulting in calcite (Anbu et al. 2016; Silva-Castro et al. 2015). In the combination of dissolution–precipitation and diffusion processes, a crystal generation process is completed:

$$6CaSO_4 + 4H_2O + 6CO_2 \rightarrow 6CaCO_3 + 4H_2S + 2S + 11O_2 \qquad (7.4)$$

Tarnishing of artistic stonework by chemical and microbial contaminants in the environment is well known (Poyatos et al. 2018). Humidity, wind, rainfall, thermal excursions, pollutant gases and dust, suspended particulate matter (SPM), and respirable suspended particulate matter (RSPM) in the environment directly influence artworks exposed to the open air (Nguyen et al. 2014). In the forms of gases and compounds, the concentration of sulfur has become increased in the atmosphere because of industrial development, suburbanization, and various other anthropogenic activities such as modern transportation, which have instigated deterioration affecting sculptures that previously remained unspoiled (Baklanov et al. 2016). The materialization of sulfuric acid due to crusts of sulfur and wet and dry deposition forms gypsum, attacking carbonate stones successively (Bonazza et al. 2009; Norvaišienė et al. 2003).

On altered stones, sulfate deposition utilized by microorganisms and its removal have been reported from various experimental observations (Dhami et al. 2014; McNamara and Mitchell 2005). Using other experiments some techniques that mandatory 16–18 days long time was directed (Heselmeyer et al. 1991). SO_2 emissions from more than 500 point sources across the world have been observed, using US National Aeronautics and Space Administration (NASA) Ozone Monitoring Instrument (OMI) satellite data (Fioletov et al. 2015). Oxidation of organic substrates associated with reduction of inorganic compounds occurs during anaerobic respiration by prokaryotes (as an analogue to respiration with O_2) to gain energy for growth (Rabus et al. 2013). Prokaryotes reduce ferric iron, manganese(IV), nitrate, elemental sulfur, sulfate, other sulfur species such as thiosulfate, and carbon dioxide, and oxidize naturally less abundant elements such as selenite, chromate (VI), arsenate(V), and uranium(VI) under anoxic conditions. In several prokaryotes under anoxic conditions the electron donor may be inorganic, which results in purely inorganic (lithotrophic) redox reactions for conservation of energy and oxidation of sulfur species with nitrate or molecular hydrogen with nitrate, iron(III), sulfate, sulfur, or CO_2 (Ghosh and Dam 2009). Dimethylsulfoxide (DMSO) and trimethylamine-N-oxide (TMAO) with some connection to inorganic electron acceptors are organic complexes (Miralles-Robledillo et al. 2019). In the presence of these compounds, the nitrogen moiety and oxygenated sulfur are reduced by anaerobic microorganisms (Pester et al. 2012). The reduction of sulfur species is most striking in anaerobic respiration because it gives rise to an exposed end product, hydrogen sulfide (H_2S), which is known as a toxic chemical with a distinctive odor (Szabo et al. 2014; Módis et al. 2014). H_2S has a conspicuous effect on the chemistry of the environment through its chemical reactivity toward iron minerals and oxygen (Chen and Morris 1972; Jørgensen 1990). Sulfide serves as an electron donor for a countless diversity of aerobic chemotrophic and anoxygenic phototrophic microorganisms, and its toxicity may form visible blooms in sulfidic habitats (Elshahed et al. 2003). The natural reduction and oxidation processes of sulfur species in the environment are known as the sulfur cycle (Burini et al. 2018). Because sulfate has thermodynamic stability and is the most abundant form of sulfur in our oxic biosphere, based on the biological sulfur cycle, sulfate is the reduction form (Labrado et al. 2019; Henrichs and Reeburgh 1987; Skyring 1987; Devereux et al. 1992).

A great diversity of sulfate-reducing microorganisms have been isolated from aquatic habitats (Vladár et al. 2008). Sulfide and its intermediates in oxidation states, such as elemental sulfur or thiosulfate, are also involved in the processes of chemical and biological oxidation (Hazeu et al. 1988; Chapman 1989) for anaerobic microorganisms that cannot reduce sulfate and serve as electron acceptors (Hedderich et al. 1998). Most frequently, sulfur-reducing anaerobic microorganisms have been isolated among these and comprised their diversity to that of sulfate-reducing microorganisms (Zhang et al. 2017; Foti et al. 2007).

This chapter discusses the reduction of sulfate or elemental sulfur by bacterial energy metabolism. The reduction of other sulfur species is also included in the growth mechanisms of bacteria. Sulfidogenic (sulfide-forming) bacteria are examples exhibiting this mechanism (Lie et al. 1999; O'Brien et al. 1984; Cord-Ruwisch et al. 1988); however, strictly speaking, it also applies to putrefying bacteria that release sulfide from sulfur-containing organic particles during their degradation (Müller and Krebs 2016).

7.2 Historical Overview

7.2.1 Sulfate-Reducing Bacteria

In aquatic habitats, the concentrations of H_2S produced is intervened biologically and the decline of sulfate recognized (Long et al. 2016). In sludge containing gypsum ($CaSO_4$) anaerobic bacterial culture oxidized to cellulose completely and used to remove sulfate (Barton and Hamilton 2007). *Spirillum desulfuricans* was the first sulfate-reducing bacterium to be sequestered and was confirmed as a strict anaerobe, producing microbial sulfide (Martins et al. 2009). Sulfur-reducing bacteria use lactate for growth and improvement, and also for cultivation (Leathen et al. 1956; Cai et al. 2010). An optimal growth temperature of 55 °C has been observed in thermophilic sulfate reduction (Vieille and Zeikus 2001), and exploitation of acetate and butyrate by a sulfate-reducing bacterium has also been reported (Rosnes et al. 1991).

In nutritional studies, *Vibrio desulfuricans* oxidizes lactate or ethanol to acetate comprehensively (Bryant et al. 1977). In addition, complete oxidation of acetate, propionate, butyrate, and other compounds to CO_2 by *Vibrio rubentschikii* has been reported (den Besten et al. 2013), and a sulfate-reducing bacterium has been reported to oxidize hydrogen (Butlin et al. 1949; O'Flaherty et al. 1998). *Clostridium nigrificans* and *Sporovibrio desulfuricans* are thermophilic, spore-forming, sulfate-reducing bacteria, which were originally described as the same species (Campbell and Postgate 1965). Cytochrome c3 production was first detected in the *Desulfovibrio* anaerobe (Rapp-Giles et al. 2000; Ozawa et al. 2000). Dissimilatory sulfate reduction biochemistry revealed differences from the pathway of assimilatory sulfate known at that time from investigations (Peck 1961; Bertran et al. 2018). In *Desulfovibrio*, adenosine-5′-phosphosulfate (APS) is not further phosphorylated to 3′-phosphoadenosine-5′-phosphosulfate (PAPS) as in the assimilatory pathway, but, to some extent, it is directly reduced to sulfite and adenosine monophosphate (AMP) (Abola et al. 1999; Peck 1961). Electron transfer has also been shown to be coupled to phosphorylation (Wilson et al. 1961). The first sulfite reductase green protein to be recognized was desulfoviridin (Laue et al. 2001; Oliveira et al. 2011).

Pyruvate and cysteine have been studied in addition to the electron acceptor sulfate in sulfate-reducing bacteria (Forsberg 1980). Incomplete oxidation of lactate, ethanol, or malate to acetate was then observed in a culture of sulfate-reducing bacteria (Nanninga and Gottschal 1987). More than 50% of organic carbon was shown to be mineralized via sulfate reduction in marine sediments (Jørgensen et al. 1990, 2019; Canfield 1989). Anaerobic enrichment studies with various organic substrates led to recognition of diverse catabolic capabilities together with degradation of aromatic organic acids in this group of microorganisms (Carmona et al. 2009; Widdel and Rabus 2001). Until the early 1980s, their nutrition, morphology, and chemical or biochemical markers (or both) were traditionally classified as those of sulfate reducers on the basis of their phenotypical characteristics (Jeanthon et al. 2002). Examples of chemotaxonomic markers are desulfoviridin (Laue et al. 2001), lipid fatty acids (Boon 1993; Ueki and Suto 1979; Taylor and Parkes 1983; Dowling et al. 1986), and menaquinones (Collins and Weddel 1986).

The sulfate-reducing bacterium *D. desulfuricans* was revealed to be related to *Myxococcus* sp. and phototrophic purple bacteria (Oyaizu and Woese 1985) on the basis of the first comparative analysis of their 16S ribosomal RNA (rRNA) sequences (Castro et al. 2000). On the bases of 16S rRNA, oligonucleotide catalogs included the spore-forming *Desulfotomaculum* sp. and various non-spore-forming sulfate-reducing bacteria were observed comprehensively (Devereux et al. 1989). Electron microscopy of the cell wall structure indicated that the *Desulfotomaculum* sp. was related to Gram-positive bacteria (Pikuta et al. 2000; Schlesner et al. 2001). The branch of Gram-negative bacteria to which all other sulfate reducers have been found to be affiliated also encompasses the sulfur-reducing *Desulfuromonas*, as well as *Myxococcus* and *Bdellovibrio* spp. (Lonergan et al. 1996; Finster et al. 1997). Sulfate-reducing microorganisms have been shown to have a unique metabolism in which five major aspects are distinguished in most research:

(a) Reduction of sulfate to sulfide, which is more complicated biochemically than O_2 reduction in aerobic organisms, involving an arrangement of enzymes (Matias et al. 2005). Sulfur may occur in eight different oxidation states, like carbon and nitrogen. In biochemistry, sulfur may form bonds to hydrogen, carbon, and oxygen, and also chains with S–S bonds. Oxidation states lower than +VI (sulfate) are rather reactive and may undergo interconversions or autoxidation even at room temperature (Buckman and Brady 1960). This complicated reactivity scrutinizes of intermediates in sulphur metabolism (Hébert et al. 2013; Filipovic et al. 2017).
(b) The wide variety of organic compounds utilized by sulfate-reducing bacteria (Cord-Ruwisch et al. 1987). Even though these are of low molecular mass and relatively simple in their structure, their oxidation under anoxic conditions often includes biochemically captivating reactions (Enning and Garrelfs 2014).
(c) The reducing equivalent flow of [H] electrons from the electron donor to the electron acceptor is associated with respiratory energy conservation, and a great variety of electron carriers seem to be involved (Lodish et al. 2000; Alberts et al. 2002).
(d) Synthesis of cell material is expected to proceed from most organic substrates via pathways commonly known from other bacteria; therefore, this has not been a major field of research (Scheffers and Pinho 2005). However, the capability of a number of sulfate-reducing bacteria for cell synthesis solely from CO_2 (and mineral salts) during growth on H_2 and with SO_4^{2-} as the sole energy source has attracted particular attention (Agostino and Rosenbaum 2018).
(e) A fifth main aspect, metabolic regulation, is largely unexplored in sulfate-reducing bacteria.

In the study of all of these aspects, molecular and genetic analyses are of increasing importance (Plugge et al. 2011).

7.2.2 Reduction of Sulfate to Sulfide

The general pathway of *Archaeoglobus* is analogous to that recognized in sulfate-reducing bacteria for sulfate reduction to sulfide (Hocking et al. 2014; Prange et al. 1999). The presence of enzymatic activities is necessary for dissimilatory reduction of sulfate (adenosine triphosphate (ATP) sulfurylase, APS reductase, and sulfite reductase) and was established in *Archaeoglobus fulgidus* (Sperling et al. 1998). In an ATP-dependent reaction to APS, sulfate is activated prior to reduction, a reaction catalyzed by ATP sulfurylase (Prioretti et al. 2014; Herrmann et al. 2014). For sulfate adenylyltransferase (sat) the coding gene was cloned, demonstrating homology with the coding genes of homo-oligomeric ATP sulfurylase from numerous bacteria and eukaryotes (Jaramillo et al. 2012). The recombinant protein was purified in *Escherichia coli* with overexpression and cloning of the *sat* gene (Baneyx 1999). Testing proved that the recombinant protein could positively form ATP from APS and pyrophosphate (PPi) (Ullrich et al. 2001). The enzyme APS reductase, in the process of APS reduction, catalyzes the two-electron reduction of APS to sulfite and AMP. *A. fulgidus* fungus has been purified and characterized (Dahl and Trüper 2001). One flavin adenine dinucleotide (FAD) and [FeS] clusters were determined and an apparent molecular mass of 160 kDa protein detected (Smutná et al. 2014). The enzyme fenced two distinct [4Fe-4S] clusters, which demonstrated similarity to those documented in APS reductase from *Desulfovibrio gigas*, established using sanitized enzymes in spectroscopic observations (Bhave et al. 2011; Oliveira et al. 2011). Molecular masses of 80 kDa and 18.5 kDa bands were exposed after study of purified APS reductase, using sodium dodecyl sulfate–polyacrylamide gel electrophoresis (SDS-PAGE) (Chiang et al. 2009). The analysis confirmed the genes coding for two various subunits a- and b-, aprA and aprB, respectively. The genes *aprA* and *aprB* encoded 73.3 kDa and 17.1 kDa polypeptides (Speich et al. 1994; Ramesh et al. 1995). The *aprA* gene product showed homologies with flavoproteins from *E. coli* and *Bacillus subtilis*, and the aprB gene contained sequences for cysteine clusters that could ligate the [FeS] centers revealed in the spectroscopic analysis (Gitt et al. 1985; Langer et al. 2013).

The reduction of sulfate to sulfide forms occurs via a number of intermediates by eight electron step practices (Toran and Harris 1989; Thom and Anderson 2008). Various nitrate-reducing bacteria and sulfate-reducing bacteria produce sulfide products directly without passing through intermediate oxidation states. Only in two cases in *D. desulfuricans* has minor concentration of sulfite or thiosulfate been observed (Akagi 1995); this does not mean that thiosulfate is a straight intermediate. In the cytoplasm or in association with the inner side of the cytoplasmic membrane, sulfate has to be taken up into the cell at all enzymatic stages from sulfate to sulfide. Sulfate uptake in sulfate-reducing bacteria is driven by an ion gradient, as has been revealed in *Desulfovibrio* sp., *Desulfobulbus propionicus*, and *Desulfococcus multivorans* (Cypionka 1987, 1994, 1995; Fitz and Cypionka 1989; Warthmann and Cypionka 1990). Instant shifts of pH in active cell suspensions on addition of sulfate in freshwater species such as *D. desulfuricans* and *D. propionicus* have revealed that the sulfate is transported simultaneously with protons (Kreke and Cypionka 1992).

Sulfate uptake in halotolerant species such as *Desulfovibrio salexigens* and *D. multivorans* involves sodium ions (Warthmann and Cypionka 1990; Stahlmann et al. 1992). With the efflux of a neutral end product, H_2S, the sulfate transport is electrogenic. The driving force is mainly the electric component of the electrochemical potential for sulfate transport and, to a lesser extent, the cation concentration gradient. There is an indication that an H^+/Na^+ antiporter creates a sodium gradient across the cytoplasmic membrane of sulfate-reducing bacteria (Gutknecht and Walter 1981; Kreke and Cypionka 1992). The sulfate uptake completely fluctuates from dissimilatory sulfate reduction for biosynthesis (assimilatory sulfate reduction). Transportation of sulfate for assimilation has been shown to occur via an ATP-binding cassette (ABC) transporter connecting to a periplasmic binding protein in *E. coli* (Hryniewicz et al. 1990; Sirko et al. 1990). In cyanobacteria and *Anacystis*, this type of mechanism is also expected for sulfate uptake (Jeanjean and Broda 1977). Sulfate uptake via ABC transporters for anabolic (assimilatory) purposes is irreversible (1 ATP/SO_4^{2-}), though this degeneracy of energy is inconsequential in relation to the comparatively small portion of concentrated sulfur required for cell synthesis (Balakrishnan et al. 2004; Gigolashvili and Kopriva 2014). The activation of the free sulfate dianion (SO_4^{2-}) with its oxygen atoms in a tetrahedral structure is slow moving chemically and not achieved effortlessly (Yant et al. 1936).

7.2.3 Sulfur-Reducing Bacteria

Electron acceptor reduction and the associated processes of electron transport and energy conservation in sulfur-reducing bacteria are the main issues, as observed in studies of *Wolinella succinogenes* (Hedderich et al. 1998). Additionally, pathways of acetate oxidation (viz., the capability for terminal oxidation of organic substrates) have been examined in *Desulfuromonas* and *Desulfurella*. In sulfur-reducing bacteria, other types of organic substrate metabolism have been of marginal research interest, in contrast to the interest in sulfate-reducing bacteria (Agostino and Rosenbaum 2018). A respiratory mechanism (such as the supplementary reduction found in numerous organisms) not coupled to sulfur reduction may arise from reactions with sulfhydryl groups of proteins or glutathione oxidized to disulfides, as shown below (Roy and Trudinger 1970):

$$2RSH + S \rightarrow RSSR + H_2S \qquad (7.5)$$

Biological reduction of sulfur to sulfide with endogenous or additional organic electron donors has been described numerous times (Liamleam and Annachhatre 2007; Burini et al. 2018). This reaction has been observed in bacteria, fungi, plants, animal tissue, and cell extracts. The processes of sulfur reduction appear to be by-reactions (incidental sulfur reduction) and were originally detected in several examples of a preciously created situation without bioenergetic or ecological importance (Gahan and Schmalenberger 2014; Masuda et al. 2016). *Desulfuromonas acetoxidans*, an obligatory mesophilic anaerobe using acetate as an electron donor, was the first pure culture definitely identified as growing by sulfur reduction (Pfennig and Biebl 1976; Peck and Odom 1981). The bacterium was discovered in a deep-green phototrophic culture initially known as *Chloropseudomonas ethylica*; as a chemotrophic partner, it was supposed that this culture was related to green sulfur bacteria, but it diverged from them by its capability to propagate on acetate and even ethanol without addition of sulfide as an electron donor (Fröstl and Overmann 2000). The actual process was explained as a sulfur sulfide series in a culture using a green phototrophic sulfur bacterium that oxidized sulfide to elemental sulfur. *Desulfuromonas* reduces sulfur with organic complexes (Okafor 2011; Ollivier et al. 2018). *Desulfuromonas*, an obligate anaerobe, was the first pure culture revealed to oxidize acetate and additional organic substrates entirely to CO_2 (Pfennig and Widdel 1981; Caccavo et al. 1996). Previously, anaerobic acetate oxidation was known only in denitrifying bacteria. These sulfur reducers were revealed to reproduce on acetate and fumarate (Yoon et al. 2013; Scholten and Stams 2000). Sulfur respiration with H_2 (Ectothiorhodospira Pelsh 1936) was established by isolation of a spirilloid bacterium (strain 5175), which, in addition, used formate (Macy et al. 1986).

Fumarate was also used as an alternative electron acceptor. Therefore, another spirilloid bacterium, which was morphologically similar, exhibiting anaerobic catabolism of fumarate (or aspartate), was documented as a facultative sulfur-reducing bacterium that oxidized H_2 or formate (Kodama and Watanabe 2003; Ouattara et al. 1992). This was a spirilloid *Campylobacter* sp. (Laanbroek et al. 1978, 1979), isolated on lactate and DMSO (Zinder and Brock 1978). Neither spirilloid sulfur reducers nor *Desulfuromonas* are capable of reducing sulfate (Hedderich et al. 1998). However, sulfur reduction has been identified in sulfate-reducing bacteria by observing and measuring their growth. In the presence of sulfur, growth on lactate or ethanol was determined in *D. gigas* (Krämer and Cypionka 1989; Badziong et al. 1978), in an isolate known as *Desulfovibrio multispirans* (Czechowski et al. 1984), and, with similar nutrition, in rod-shaped, desulfoviridin-negative sulfate reducers related to the proposed genus *Desulfomicrobium* (Nazina et al. 1985).

Isolated ferric iron–reducing anaerobic bacteria have been shown to be facultative sulfur reducers (Balashova 1985; Myers and Nealson 1988a, b). Sulfur-reducing *Desulfuromonas* have also been shown to reduce ferric iron (Roden and Lovley 1993). Moreover, a *Pelobacter* sp. isolated as a fermentative bacterium has been documented as a facultative reducer of sulfur and ferric iron (Lovley et al. 1995). A sulfur-reducing, acetate-oxidizing anaerobe, *Desulfurella acetivorans* (Bonch-Osmolovskaya et al. 1990), is moderately thermophilic. Moreover, *Aquifex* (Huber et al. 1992), *Ammonifex* (Huber et al. 1996), and *Desulfurobacterium* (L'Haridon et al. 1998) have been designated as hydrogen-utilizing, sulfur-reducing bacteria. *Ammonifex* was previously isolated as a nitrate-reducing bacterium. Sulfur-reducing bacteria very actively metabolized acetate, certain redox proteins, and metal centers

(Probst et al. 1977; Zöphel et al. 1988) in the first biochemical studies. The citric acid cycle was shown to occur by acetate oxidation, either with preliminary activation by coenzyme A (CoA) transfer from succinyl-CoA, as in *Desulfuromonas* (Gebhardt et al. 1985), or with ATP-dependent acetate activation, as in *Desulfurella* (Galushko and Schink 2000).

7.2.4 Sulfur-Reducing Archaea

Extremely thermoacidophilic microorganisms were first described in the early 1970s (Brock et al. 1972; Brierley and Brierley 1982). Aerobic sulfur oxidizer *Sulfolobus* species were the organisms classified. After some time, they were documented as members of a new kingdom of life, termed Archaebacteria (Woese and Fox 1977; Woese et al. 1978). Extremely thermophilic novel archaea that grew anaerobically and produced sulfide were identified (Fischer and Tillmanns 1988), and the number of novel isolates gradually increased each year. Various isolates that reduced sulfur in a by-reaction or using an electron sink to facilitate fermentation were described (Zillig et al. 1982). Sulfur respiration as a mode of energy metabolism in archaea was clearly demonstrated in cultures of *Thermoproteus* and *Pyrodictium* spp. in which H_2 was the only electron donor in the absence of organic compounds (Fiala and Stetter 1986). Additionally, newly isolated archaea that certainly grew by sulfur respiration (viz., using H_2 and sulfur) were *Stygioglobus azoricus* (Segerer et al. 1991), *Pyrobaculum islandicum* (Huber et al. 2000), and *Stetteria hydrogenophila. Acidianus infernus* and *Desulfurolobus ambivalens* are newly isolated lithoautotrophic thermophiles that grow aerobically with unique adaptability of sulfur metabolism through sulfur oxidation, as well as anaerobically through sulfur reduction with H_2 (Segerer et al. 1985, 1986).

The hydroxypropionate pathway and the reductive citric acid cycle have been shown to be operative in *Thermoproteus neutrophilus* and *Acidianus*, respectively, in carbon assimilation during sulfur reduction with H_2 (Schäfer et al. 1986; Menendez et al. 1999). In several hyperthermophiles, sugar metabolism pathways have also been investigated, such as the sulfur-respiring, facultatively organotrophic *Thermoproteus tenax* (Siebers and Hensel 1993; Selig et al. 1997). Evidence has also been provided for a nonphosphorylated Entner–Doudoroff pathway and an altered Embden–Meyerhof-pathway (Wolf et al. 2016).

Moreover, in the facultatively organotrophic sulfur-respiring species *T. tenax* and *P. islandicum*, complete oxidation of organic substrates via the citric acid cycle has been demonstrated (Selig and Schönheit 1994). In *Pyrodictium brockii* and *Pyrodictium abyssii*, electron transport has been studied during sulfur reduction with H_2, and these species employ altered transport chains (Pihl et al. 1992; Hao and Ma 2003).

It is well known that in the atmosphere, the presence of chemical and microbial pollutants accelerates degradation of artistic stonework (Ciferri et al. 2000). These phenomena are directly influenced by the environment; humidity, wind, rainfall, thermal excursions, pollutant gases, and dust are especially hard on artwork exposed to the open air (Didenko and Skripnuk 2014; Saiz-Jimenez 1993). In the atmosphere, nitrogen, sulfur, carbon oxide, and pollutants have increased as a result of urbanization, transport, and industrial activity, causing a substantial acceleration in degradation of artwork that previously remained well preserved (Chen et al. 2007; Devalia et al. 1994). Wet and dry deposition of sulfated crusts in the presence of sulfuric acid play a significant role in the degradation, attacking carbonate stones with consequent gypsum materialization (Sáiz-Jiménez et al. 1991; Bonazza et al. 2009). To remove sulfate deposition on altered stones, microorganisms have been utilized in some investigations (Heselmeyer et al. 1991). Recently, research on deployment of prudently selected microbial cultures developed on an appropriate type of support has been instigated with the aim of developing and improving a system of biological removal of sulfates present on artistic stonework (Soffritti et al. 2019).

7.2.5 Sulfate-Reducing Archaea

The sulfate-reducing archaeon *A. fulgidus* was isolated from a submarine hydrothermal area and recognized as the first representative of the archaeal domain of life that could preserve energy via dissimilatory sulfate reduction (Burggraf et al. 1990a, b; Zellner et al. 1998). Two other archaeal species, *Archaeoglobus profundus* and *Archaeoglobus lithotrophicus*, are also sulfate reducers (Burggraf et al. 1990a, b; Stetter et al. 1993). *A. veneficus* is the fourth *Archaeoglobus* species to be identified and uses sulfite, but not sulfate, as an electron acceptor (Steinsbu et al. 2010). *Archaeoglobus* species are typically cultivated at a temperature above 80 °C and require at least 10 grams of NaCl per liter for growth. Thiosulfate is used as an electron acceptor for oxidation of H_2 by another hyperthermophilic archaeon known as *Ferroglobus placidus* (Manzella et al. 2015; Hocking et al. 2014). For complete oxidation of lactate to CO_2, the pathway could be interpreted as carbon metabolism. The CO dehydrogenase pathway in mesophilic sulfate-reducing bacteria has been recognized as an archaeal parallel pathway (Schauder et al. 1987; Schauder et al. 1988), with envelopment of archaeal cofactors (Möller-Zinkhan et al. 1989; Schmitz and Baeuerle 1991; Klein et al. 1993). *A. lithotrophicus* has been recognized to use a reductive CO dehydrogenase pathway in autotrophic CO_2 fixation, and similar CO_2 incorporation occurs in sulfate-reducing bacteria (Vorholt et al. 1997; Hügler et al. 2005). This sulfate reduction process has been shown to include the same enzymatic phases as those seen in other bacterial sulfate reducers. At the gene level, enzymes involved in the sulfate reduction pathway have been distilled and equated to similar bacterial enzymes in *Archaeoglobus* (Basen et al. 2011; Bertran et al. 2018). The first complete genome sequence of this type to be published was that of the sulfate-reducing prokaryote *A. fulgidus* in 1997 (Klenk et al. 1997).

7.3 Microorganisms Reducing Sulfur Compounds Other Than Sulfate or Sulfur

Dissimilatory reduction of sulfur compounds other than sulfate and sulfur (especially sulfite and thiosulfate) has been regularly observed among sulfate-reducing microorganisms (Widdel and Bak 1992; Odom and Singleton 1993). Various observations involving tetrathionate and thiosulfate reduction in bacteria other than sulfate or sulfur reducers have been reported (Barrett and Clark 1987); these prokaryotes utilize sulfur compounds as electron acceptors but reduce neither sulfate nor elemental sulfur. Thiosulfate is the reduction product of tetrathionate; reduction occurs in these sulfur species, especially among enterobacteria (Hazeu et al. 1988; Balci et al. 2017). Often, the capabilities to reduce tetrathionate and thiosulfate appear together, as has been elucidated using one reductase for both compounds (Oltmann and Stouthamer 1975; Oltmann et al. 1975). The presence of an electron transport chain to tetrathionate, permitting respiratory energy conservation in *Citrobacter* and *Proteus*, was substantiated in a study in which the growth substrates were sugars (Novotný and Kapralek 1979). The thiosulfate that is formed may be further reduced to sulfide and sulfite, the latter often being an end product that is not reduced further. In *Clostridium pasteurianum*, sulfite reductase has been induced by sulfite and distinguished from the assimilatory enzyme (Barrett and Clark 1987; Laue et al. 2001). Reduction of thiosulfate to sulfite and sulfide catalyzed by yeast cells has been described (Akagi 1995; Peck 1960; de Vito and Dreyfuss 1964).

7.4 Biological Methods for Control of Microbial Contamination

Nowadays, biorestoration technologies including bioconsolidation and biocleaning of inorganic and organic deposits are being widely applied to cultural heritage objects (Ranalli et al. 2005; Bosch-Roig and Ranalli 2014). Biorestoration methods, which exploit altered strains of sustainable bacteria, provide useful alternatives to traditional and more invasive chemical or mechanical methods for resolution of various types of stone destruction (Joutey et al. 2013; Jain et al. 2005). Thus far, bacteria have been mostly used for their capability to metabolize or produce inorganic compounds in curative concentrations, and various observations have been carried out to evaluate their use as possible neutralizing agents on artwork (Haas and Franz 2009). Application of parasitic and antagonistic species of biodeteriogens and products of secondary metabolism of microorganisms and/or plants—known as green biocides—are currently being studied for use in biological restoration methods (Rogerio-Candelera et al. 2013).

For artwork restoration, biocide treatments are traditional and widely used methods. *B. subtilis* and *Bacillus amyloliquefaciens* have capabilities to produce antifungal peptides (Sutyak et al. 2008). Antifungal peptides, antifungal lipopeptides, and antimicrobial polypeptides are known as bioactive peptides (Matejuk et al. 2010). Through defense mechanisms in stress conditions, microbial lipopeptide compounds are produced naturally. Lipopeptides such as biosurfactants produced by *Bacillus* sp. inhibit the growth of *Cladosporium* spp., *Penicillium* spp., *Fusarium oxysporum* and *Aspergillus niger* fungi isolated from biodegraded mural paintings (Velho et al. 2011; Silva et al. 2015). Through analysis of a microbial population contaminating a seventeenth-century easel painting ascribed to the painter Carlo Bononi, it was found that a *Bacillus*-based compound (including *B. subtilis*, *B. pumilus*, and *B. megaterium*) had the potential to prevent biodeterioration of artworks (Santoro et al. 2014; Caselli et al. 2018).

According to several reports, the antimicrobial activity of *Bacillus* spp. has the ability to counteract various types of pathogenic bacteria (Chalasani et al. 2015). *Bacillus* spores are utilized as antifungal agents in agriculture, aquaculture, and zootechnics, and have a long history of safe use in humans (Leyva Salas et al. 2017). These uses are mainly based on the enzymatic activity and competitive antagonism of *Bacillus*, although the manufacture of bacteriocins and other bioactive compounds might also be convoluted (Noreña and Rigon 2018; Muñoz-Dorado et al. 2016). *Bacillus* spores added to detergent solution cleaning products for treatment of the biological patina on stone materials showed no interaction with stone samples (absence of decohesion, discoloration, fractures, or porosity increases) and have thus been suggested as potential biological agents for biorestoration in the field of cultural heritage conservation (Venkateswaran et al. 2004). Studies investigating the use and applicability of biological systems for biorestoration of stonework are summarized in Table 7.1.

7.5 Energy Metabolism of Sulfur-Reducing Bacteria

7.5.1 Pathways of Sulfate Reduction

Two sulfate reduction biological pathways have been reported. The first is the assimilatory pathway, which is widespread in the three domains of life; in small amounts, sulfate is reduced to sulfide and transformed into cysteine, of which added biological sulfur-holding molecules are derivatives (Grein et al. 2013; Günal et al. 2019). In the dissimilatory pathway, which is constrained to five bacterial and two archaeal heredities, sulfate is the terminal electron acceptor of the respiratory pathway, producing large amounts of sulfide (Vimr et al. 2004; Fewson 1988; Louis and Flint, 2009; Pace 1997). Sulfite is reduced directly to sulfide, and the creation of a mixture of products (as well as trithionate and thiosulfate) in relative quantities that are determined by the reaction conditions is a key difference between the assimilatory and dissimilatory pathways (Oliveira et al. 2008; Leavitt et al. 2015). The activation of sulfate starts two pathways (Fig. 7.1) through a reaction with ATP to form APS, a phase catalyzed by the trimeric sulfate adenylyltransferase (Sat), also known as ATP sulfurylase (Sperling et al. 1998; Grein et al. 2013).

APS formation is endergonic and is determined by hydrolysis of pyrophosphate by pyrophosphatase (soluble or membrane bound). Therefore, the two ATP consumed for the activation of sulfate to APS (Shen and Buick 2004; Barton and

Table 7.1 Biorestoration compounds and methods reported in the scientific literature for removal of sulfur from stonework

Microorganisms	Application	Mechanism	Reference
Aerobacter aerogenes, *Comamonas* spp., *Pseudomonas stutzeri*	Application with cotton wool layer/agar carrier	Biocleaning of graffiti paint from granite and concrete surfaces (in in vitro tests)	Sanmartín and Bosch-Roig (2019)
Desulfovibrio desulfuricans	Immersion	Removal of sulfates from marble (in in vitro tests)	Atlas et al. (1988)
Desulfovibrio desulfuricans	Immersion	Removal of sulfates from marble (in in vitro tests)	Gauri et al. (1992)
Desulfovibrio spp.	Use of sepiolite as a delivery system	Removal of sulfates from marble (in in vitro tests)	Ranalli et al. (1997)
Desulfovibrio vulgaris subsp. *vulgaris*	Use of sepiolite/ hydrobiogel-97/ Carbogel as a carrier	Removal of black crusts from marble	Cappitelli et al. (2006)
Desulfovibrio vulgaris subsp. *vulgaris*, *Pseudomonas pseudoalcaligenes*, *Pseudomonas stutzeri*	Use of Carbogel/ mortar and alginate beads	Removal of sulfates from sandstone walls	May et al. (2008)
Desulfovibrio vulgaris subsp. *vulgaris*, *Pseudomonas pseudoalcaligenes*	Use of Carbogel with a multilayer biosystem	Removal of sulfates from sandstone walls	Alfano et al. (2011)
Desulfovibrio vulgaris, *Desulfovibrio desulfuricans*, *Desulfurivibrio alkaliphilus*	Use of sepiolite	Removal of sulfates from sandstone walls	Ranalli et al. (2000)

Hamilton 2007). APS is converted to PAPS by adenylyl-sulfate kinase (CysC) in the prokaryotic assimilatory pathway, thioredoxin-dependent PAPS reductase (CysH) reduces PAPS to sulfite, and, finally, sulfite is reduced to sulfide by an assimilatory sulfite reductase that is multimeric and NADPH [reduced nicotinamide adenine dinucleotide phosphate] dependent (CysIJ) or a monomeric ferredoxin-dependent enzyme (Chartron et al. 2007; Schmidt and Jäger 1992; Campanini et al. 2015). APS reductase (AprBA), a heterodimeric iron–sulfur flavoenzyme, reduces APS to sulfite in the dissimilatory pathway. The dissimilatory sulfite reductase DsrAB, a siroheme containing protein, reduces sulfite with a contribution from the small protein DsrC (Oliveira et al. 2008; Meyer and Kuever 2008).

The majority of DsrC is not associated with DsrAB and is thus free to interact with other proteins. DsrC was initially believed to be a third subunit of DsrAB (Santos et al. 2015; Oliveira et al. 2011). DsrD is a small protein, which is often encoded downstream of dsrAB, and might also be involved in sulfite reduction, probably in a regulatory role, but its particular function is still unknown (Mussmann et al. 2005; Arendsen et al. 1993). Remarkably, the dsrD gene is strongly downregulated in the presence of high sulfide concentrations (Aherne et al. 2012;

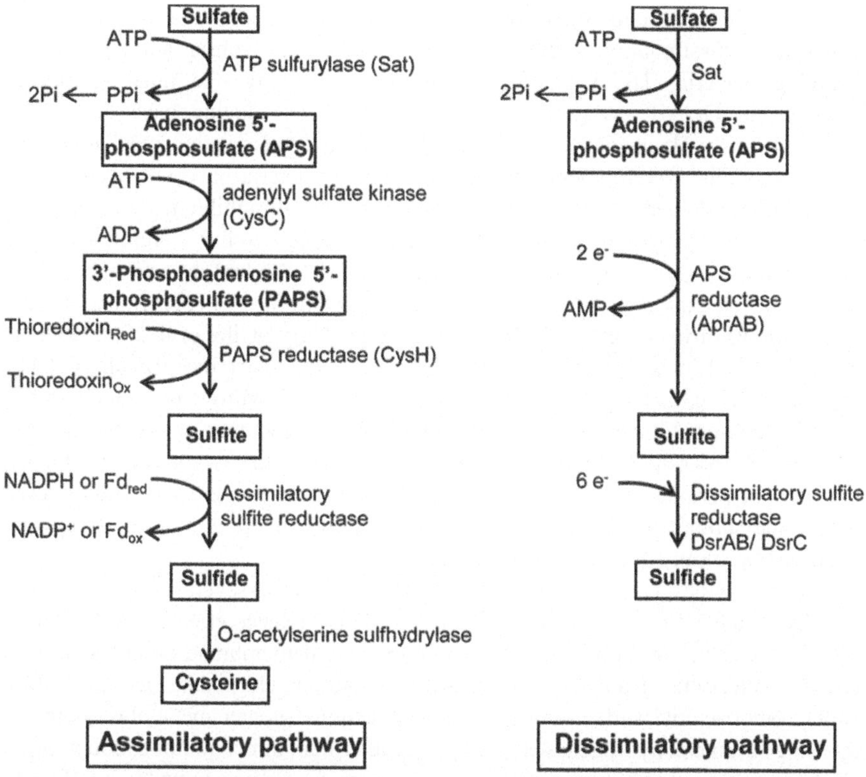

Fig. 7.1 Prokaryotic assimilatory and dissimilatory pathways of sulfate reduction (adapted from Grein et al. 2013)

Caffrey et al. 2012). An α_2–β_2 unit is formed by DsrAB sulfite reductase, which contains two siroheme cofactors per α–β unit, coupled to a [4Fe–4S] iron–sulfur cluster through the cysteine heme axial ligand (Simon and Kroneck 2013; Parey et al. 2013) in which only one of the cofactors is catalytically active. They include a ferredoxin domain, and the DsrA/DsrB proteins also have a modular character, possibly acting as the electron donor to a precursor enzyme that was previously integrated into the reductase gene sequence (Schiffer et al. 2008; Dahl 2008; Myers and Nealson 1988a, b). The dsrA and dsrB genes are paralogous and seem to be consequent from a gene duplication event preceding the differentiation of the Archaea and Bacteria domains (Makarova et al. 2005). Moreover, the assimilatory sulfite/nitrite reductases also show an internal twofold symmetry of a module related to DsrA/DsrB, suggesting that they also resulted from a gene duplication event (Oliveira et al. 2008; Grein et al. 2013). The mutual origin of the assimilatory and dissimilatory enzymes from an ancestral gene that was present in one of the initial life-forms on earth is confirmed by these data (Dhillon et al. 2003; Berg et al. 2010).

The respiratory membrane complexes QmoABC and DsrMKJOP play an important role in dissimilatory sulfur metabolism and energy conservation in sulfate-reducing bacteria. The physiological partners of the two terminal reductases AprBA and DsrAB are strictly conserved in sulfate-reducing bacteria (Pereira et al. 2011a, b; Hausmann et al. 2018). The energy conservation mechanisms of these membrane developments in sulfate-reducing bacteria have not been clearly identified (Heidelberg et al. 2004). An abundant pool of multiheme *c* cytochromes have been categorized in *Desulfovibrio* and other sulfate-reducing deltaproteobacteria (Tian et al. 2017). In an anaerobe the first cytochrome c to be described was tetraheme cytochrome *c*3 (type I cytochrome *c*3 (TpIc3)), and it is the most widely conveyed protein in *Desulfovibrio* spp. (Rapp-Giles et al. 2000; Valente et al. 2001). Either formate and hydrogen, or both, assist the growth of these organisms. TpIc3 is the periplasmic electron acceptor of hydrogenases and formate dehydrogenases (Venceslau et al. 2010, 2014). In the genomes of sulfate-reducing bacteria, the presence of the cycA gene coding for TpIc3 is correlated with the presence of periplasmic hydrogenases and formate dehydrogenases, which lack a membrane subunit for direct reduction of quinone (Venceslau et al. 2010; Pereira et al. 2011a, b), in contrast to most bacteria.

These enzymes have an active cytochrome *c*3 subunit in several cases. A proton-coupled two-electron transfer is performed by TpIc3 (Reece et al. 2006; Costentin 2008). It can deliver this reducing power to several membrane complexes, as it receives electrons from H_2 or formate oxidation (Jormakka et al. 2002). Deltaproteobacterial sulfate-reducing bacteria acquire greater metabolic flexibility by using soluble dehydrogenases and TpIc3, rather than direct quinone reduction, as electrons can be shuttled from side to side in numerous alternate pathways (Duarte et al. 2018; Plugge et al. 2011). Deltaproteobacteria sulfate-reducing bacteria may derive additional electrons from intracellular cycling of redox intermediates such as hydrogen and formate, in contrast to other groups of sulfate-reducing bacteria (Junier et al. 2010; da Silva et al. 2013). *Geobacter*, *Shewanella*, *Anaeromyxobacter*, and *Desulfovibrio* are exposed to variable redox circumstances, and high content of multiheme c cytochromes seems to be characteristic of soil and sediment proteobacteria (Newsome et al. 2014; Majumder and Wall 2017).

7.6 Biorestoration Using Microorganisms as Restorative Agents

New restoration methods now being implemented involve application of advantageous microorganisms as restorative agents that are safe for human and environmental health (Biondi et al. 2002; Tilman et al. 2002). Recovery and conservation of artworks through use of such innovative biotechnologies are proposed, using the enzymatic/metabolic activities of the microorganisms (Soffritti et al. 2019). Isolation, screening, and selection of microorganisms have been done using environment matrices to select those with the capability for biocleaning and biocalcification, for use in bioconsolidation of stonework and monuments, or as potential decontamination agents (Schillinger et al. 1996; Wilmes and Bond 2004). Suitable and nonpathogenic microorganisms can be identified using traditional microbiological approaches because of the high level of microbial natural biodiversity, without the need to employ genetically modified organisms (GMOs) (Oliver 2014). It has been observed that biorestoration can be used as an effective substitute for traditional restoration approaches (Urbina et al. 2019). Biological methods can be considered sustainable on the basis of the following features: (1) exploitation of natural microbial procedures, as bacterial activity leads to natural release of carbon dioxide and water, so their engagement can be considered ecologically viable; (2) selection of nonpathogenic microorganisms that are safe to use, with no recognized hazard potential; (3) environmental friendliness due to absence of the hazardous waste and emissions associated with chemicals approaches; (4) gentle action on artworks; and (5) lower or similar costs in comparison with conventional methods.

7.7 Bioremediation and Bioconsolidation

Use of various enzymes such as lipases, proteases, and carbohydrates with biocalcifying bacteria, sulfate-reducing bacteria, nitrate-reducing bacteria, and hydrolytic activity are central to bioremediation (Ayangbenro et al. 2018). For consolidation and reinforcement of stone, biocalcifying bacteria are used, whereas for sulfate and nitrate reduction, sulfate- and nitrate-reducing bacteria are employed (Cappitelli et al. 2006; Balloi et al. 2017). Sulfate-reducing bacteria reduce sulfate to gaseous hydrogen sulfide, and nitrate-reducing bacteria reduce nitrates to gaseous nitrogen or nitrous oxide, in particular (Fida et al. 2018). Additionally, the hydrolytic activities of microbes are applicable for elimination of organic deposits from artworks, such as carbohydrate, proteinaceous, and lipidic residues (Guimaraes 2012; Soffritti et al. 2019). Bioconsolidation involves resolution or reversal of deterioration of the calcareous medium of stonework, which is influenced by natural weathering processes and the climatic–ecological conditions to which the artworks are exposed (Jroundi et al. 2012; Jiménez-López et al. 2007). Some bacteria cause calcite precipitation due to metabolic activity with active and passive mechanisms.

Bacterial activity encourages chemical changes in the microenvironment through passive carbonatogenesis, which leads to carbonate and bicarbonate ion accumulation and to solid particle precipitation (Zhu and Dittrich 2016). In several metabolic pathways of the sulfur and nitrogen cycles, passive precipitation is induced (Castanier et al. 2000; Hammes and Verstraete 2002). Carbonate particles are produced by ionic exchange through the cell membrane in active precipitation (Al-Thawadi 2011; Melim Kristen et al. 2001; Northup et al. 2001; Cacchio et al. 2003). *Bacillus pasteurii*, *B. subtilis*, *Myxococcus xanthus*, *Acinetobacter*, *Pseudomonas*, *Pantoea*, and *Cupriavidus* strains have long been established to be effective in calcite precipitation for preservation purposes and are widely considered for biomineralization (Daskalakis et al. 2013; Rodriguez-Navarro et al. 2003; Chafetz 1986; Warthmann et al. 2000).

7.8 Biocleaning

Artwork restoration has involved biological cleaning since the early 1990s, and nowadays this technology is proficiently used for recovery and conservation of artworks and monuments, limiting the need for use of conventional chemical products (Bosch-Roig and Ranalli 2014; Sterflinger and Piñar 2013). Different viable bacterial cultures employed in biocleaning have been used for recovery of numerous signs of deterioration (black crusts, mineral salt deposits, and organic matter residues) and have been shown to be more useful than chemical approaches in eliminating complex and enveloped substrates, thanks to the extremely selective and specific enzymatic actions of microbes (Ye et al. 2016; Cappitelli 2016). For instance, disfiguring black crusts on stonework consist of a mixture of calcium sulfate dihydrate (gypsum) deposits on the stone and soot particles resulting from air pollution (Warscheid and Braams 2000; Ortega-Calvo et al. 1995). The first microorganism to be appraised for use as a cleaning agent for reduction of gypsum and black crusts on marble was the anaerobic sulfate-reducing bacterium *D. desulfuricans*. Since that pioneering work, *D. desulfuricans* and *Desulfovibrio vulgaris* have been effectively used in bioremediation of historic stonework and monuments (Gilmour et al. 2011; Ayangbenro et al. 2018).

7.9 Conclusion and Future Prospects

Demarcated metabolic capability or particular nutrient requirements result from microorganism selection. Enrichment of stones with nitrogen and sulfur compounds opens the way for colonization by sulfur-oxidizing and nitrifying bacteria. For balanced growth and carbohydrate sheath production, cyanobacteria need sulfates to find a good niche in weathered stone with gypsum crusts. Construction of biofilms whose extracellular polymers are firmly bound to inert surfaces is habitual among microorganisms. The impact of pollutants in biotransformation is exacerbated by accumulation of microorganisms on stone surfaces. Particularly in slow-growing microorganisms, biofilm growth is evidence of greater accumulation of cells. Within the depths of an aerobic biofilm, an anaerobic microenvironment can be produced, and the sulfate-reducing bacteria *D. desulfuricans* can be actively involved in transformation of gypsum. The cyanobacterium *Gloeothece* can also use gypsum for growth under aerobic conditions. Biofilm organisms may also be involved in immobilization of heavy metals. Although we are still far from a complete understanding of the physiological diversity of biofilm organisms and their interactions on the surfaces of stone monuments, biotransformation is envisaged to be an important process balanced with deposition of pollutants.

Acknowledgements The authors are greatful to Prof. Jameel Ahmad, Principal, Gandhi Faiz-E-Aam College, Shahjahanpur, Uttar Pradesh for their suggestion and constant encouragement.

References

Abola AP, Willits MG, Wang RC, Long SR (1999) Reduction of adenosine-5′-phosphosulfate instead of 3′-phosphoadenosine-5′-phosphosulfate in cysteine biosynthesis by *Rhizobium meliloti* and other members of the family Rhizobiaceae. J Bacteriol 181:5280–5287

Agostino V, Rosenbaum MA (2018) Sulfate-reducing electroautotrophs and their applications in bioelectrochemical systems. Front Energy Res 6:55

Aherne M, Lyons JA, Caffrey M (2012) A fast, simple and robust protocol for growing crystals in the lipidic cubic phase. J Appl Crystallogr 45:1330–1333

Akagi JM (1995) Respiratory sulfate reduction. In: Sulfate-reducing bacteria. Springer, Boston, pp 89–111

Alberts B, Johnson A, Lewis J, Raff M, Roberts K, Walter P (2002) Electron-transport chains and their proton pumps. Mol Biol Cell 91:401

Alfano G, Lustrato G, Belli C, Zanardini E, Cappitelli F, Mello E, Ranalli G (2011) The bioremoval of nitrate and sulfate alterations on artistic stonework: the case-study of Matera Cathedral after six years from the treatment. Int Biodeterior Biodegrad 65:1004–1011

Al-Thawadi SM (2011) Ureolytic bacteria and calcium carbonate formation as a mechanism of strength enhancement of sand. J Adv Sci Eng Res 1:98–114

Anbu P, Kang CH, Shin YJ, So JS (2016) Formations of calcium carbonate minerals by bacteria and its multiple applications. Springer Plus 5:1–26

Arendsen AF, Verhagen MF, Wolbert RB, Pierik AJ, Stams AJ, Jetten MS, Hagen WR (1993) The dissimilatory sulfite reductase from *Desulfosarcina variabilis* is a desulforubidin containing uncoupled metalated sirohemes and S = 9/2 iron–sulfur clusters. Biochemist 32:10323–10330

Atlas RM, Chowdhury AN, Gauri KL (1988) Microbial calcification of gypsum-rock and sulfated marble. Stud Conserv 33:149–153

Ayangbenro AS, Olanrewaju OS, Babalola OO (2018) Sulfate-reducing bacteria as an effective tool for sustainable acid mine bioremediation. Front Microbiol 9:1–15

Badziong W, Thauer RK, Zeikus JG (1978) Isolation and characterization of *Desulfovibrio* growing on hydrogen plus sulfate as the sole energy source. Arch Microbiol 116:41–49

Baklanov A, Molina LT, Gauss M (2016) Megacities, air quality and climate. Atmos Environ 126:235–249

Balakrishnan L, Venter H, Shilling RA, van Veen HW (2004) Reversible transport by the ATP-binding cassette multidrug export pump LMRA ATP synthesis at the expense of downhill ethidium uptake. J Biol Chem 279:11273–11280

Balashova VV (1985) Use of molecular sulfur as an agent oxidizing H_2 by a facultative anaerobic *Pseudomonas* strain. Mikrobiologiia 54:324–326

Balci N, Brunner B, Turchyn AV (2017) Tetrathionate and elemental sulfur shape the isotope composition of sulfate in acid mine drainage. Front Microbiol 8:1564

Balloi A, Lombardi E, Troiano F, Polo A, Capitelli F, Gulotta D, Daffonchio D (2017) Sulfate reducing bacteria as bio-cleaning agents: development of new methodologies and study cases. Conserv Sci Cultural Herit 15:109–119

Baneyx F (1999) Recombinant protein expression in *Escherichia coli*. Curr Opin Biotechnol 10:411–421

Barrett EL, Clark MA (1987) Tetrathionate reduction and production of hydrogen sulfide from thiosulfate. Microbiol Rev 51:192

Barton LL, Hamilton WA (eds) (2007) Sulphate-reducing bacteria: environmental and engineered systems. Cambridge University Press, New York

Basen M, Krüger M, Milucka J, Kuever J, Kahnt J, Grundmann O, Shima S (2011) Bacterial enzymes for dissimilatory sulfate reduction in a marine microbial mat (Black Sea) mediating anaerobic oxidation of methane. Environ Microbiol 13:1370–1379

Berg IA, Kockelkorn D, Ramos-Vera WH, Say RF, Zarzycki J, Hügler M, Fuchs G (2010) Autotrophic carbon fixation in archaea. Nat Rev Microbiol 8:447

Bertran E, Leavitt WD, Pellerin A, Zane GM, Wall JD, Halevy I, Johnston DT (2018) Deconstructing the dissimilatory sulfate reduction pathway: isotope fractionation of a mutant unable of growth on sulfate. Front Microbiol 9:3110

Bhave DP, Hong JA, Lee M, Jiang W, Krebs C, Carroll KS (2011) Spectroscopic studies on the [4Fe–4S] cluster in adenosine 5′-phosphosulfate reductase from *Mycobacterium tuberculosis*. J Biol Chem 286:1216–1226

Biondi V, Iraldo F, Meredith S (2002) Achieving sustainability through environmental innovation: the role of SMEs. Int J Technol Manag 24:612–626

Bonazza A, Messina P, Sabbioni C, Grossi CM, Brimblecombe P (2009) Mapping the impact of climate change on surface recession of carbonate buildings in Europe. Sci Total Environ 407:2039–2050

Bonch-Osmolovskaya EA, Sokolova TG, Kostrikina NA, Zavarzin GA (1990) *Desulfurella acetivorans* gen. nov. and sp. nov. a new thermophilic sulfur-reducing eubacterium. Arch Microbiol 153:151–155

Boon T (1993) Tumor antigens recognized by cytolytic T lymphocytes: present perspectives for specific immunotherapy. Int J Cancer 54:177–180

Bosch-Roig P, Ranalli G (2014) The safety of biocleaning technologies for cultural heritage. Front Microbiol 5:155

Brady NC, Weil RR (1999) Nitrogen and sulfur economy of soils. Nat Prop Soil:491–523

Brierley CL, Brierley JA (1982) Anaerobic reduction of molybdenum by *Sulfolobus* species. Zentralblatt für Bakteriologie Mikrobiologie und Hygiene: I. Abt. Originale C: Allgemeine, angewandte und ökologische Mikrobiologie 3:289–294

Brock A, Raja PKS, Vise JB (1972) The palaeomagnetism of the Kisii Series, western Kenya. Geophys J Int 28:129–137

Bryant MP, Campbell LL, Reddy CA, Crabill MR (1977) Growth of *Desulfovibrio* in lactate or ethanol media low in sulfate in association with H_2-utilizing methanogenic bacteria. Appl Environ Microbiol 33:1162–1169

Buckman HO, Brady NC (1960) The nature and properties of soils. Soil Sci 90:212

Burggraf S, Fricke H, Neuner A, Kristjansson J, Rouvier P, Mandelco L, Stetter KO (1990a) *Methanococcus igneus* sp. nov., a novel hyperthermophilic methanogen from a shallow submarine hydrothermal system. Syst Appl Microbiol 13:263–269

Burggraf S, Jannasch HW, Nicolaus B, Stetter KO (1990b) *Archaeoglobus profundus* sp. nov., represents a new species within the sulfate-reducing archaebacteria. Syst App. Microbiology 13:24–28

Burini RC, Kano HT, Yu YM (2018) The life evolution on the sulfur cycle: from ancient elemental sulfur reduction and sulfide oxidation to the contemporary thiol-redox challenges. Glutathione in health and disease. Pinar Erkekoglu and Belma Kocer-Gumusel, IntechOpen, Hyattsville. https://doi.org/10.5772/intechopen.76749

Butlin KR, Adams ME, Thomas M (1949) The isolation and cultivation of sulphate-reducing bacteria. Microbiology 3:46–59

Caccavo F Jr, Coates JD, Rossello-Mora RA, Ludwig W, Schleifer KH, Lovley DR, McInerne MJ (1996) *Geovibrio ferrireducens*, a phylogenetically distinct dissimilatory Fe(III)-reducing bacterium. Arch Microbiol 165:370–376

Cacchio P, Ercole C, Cappuccio G, Lepidi A (2003) Calcium carbonate precipitation by bacterial strains isolated from a limestone cave and from a loamy soil. Geomicrobiol J 20:85–98

Caffrey C, Sengupta M, Park-Lee E et al (2012) Residents living in residential care facilities: United States, 2010. NCHS data brief, no. 91. National Center for Health Statistics, Hyattsville

Cai J, Jiang J, Zheng P (2010) Isolation and identification of bacteria responsible for simultaneous anaerobic ammonium and sulfate removal. Sci China Chem 53(3):645–650

Campanini B, Pieroni M, Raboni S, Bettati S, Benoni R, Pecchini C, Mozzarelli A (2015) Inhibitors of the sulfur assimilation pathway in bacterial pathogens as enhancers of antibiotic therapy. Curr Med Chem 22:187–213

Campbell LL, Postgate JR (1965) Classification of the spore-forming sulfate-reducing bacteria. Bacteriol Rev 29:359

Canfield DE (1989) Sulfate reduction and oxic respiration in marine sediments: implications for organic carbon preservation in euxinic environments. Deep Sea Research Part A. Oceanographic Res Papers 36:121–138

Cappitelli F (2016) Biocleaning of cultural heritage surfaces. Open Conf Proc J 7 (suppl 1: M6):65–69

Cappitelli F, Zanardini E, Ranalli G, Mello E, Daffonchio D, Sorlini C (2006) Improved methodology for bio removal of black crusts on historical stone artworks by use of sulfate-reducing bacteria. Appl Environ Microbiol 72:3733–3737

Carmona M, Zamarro MT, Blázquez B, Durante-Rodríguez G, Juárez JF, Valderrama JA, Díaz E (2009) Anaerobic catabolism of aromatic compounds: a genetic and genomic view. Microbiol Mol Biol Rev 73:71–133

Caselli E, Pancaldi S, Baldisserotto C, Petrucci F, Impallaria A, Volpe L, Bevilacqua F (2018) Characterization of biodegradation in a 17th century easel painting and potential for a biological approach. PLoS One 13:e0207630

Castanier S, Le Metayer-Levrel G, Perthuisot J P (2000) Bacterial roles in the precipitation of carbonate minerals. In: Microbial sediments. Springer, Berlin/Heidelberg, pp 32–39

Castro HF, Williams NH, Ogram A (2000) Phylogeny of sulfate-reducing bacteria. FEMS Microbiol Ecol 31:1–9

Chafetz HS (1986) Marine peloids; a product of bacterially induced precipitation of calcite. J Sediment Res 56:812–817

Chalasani AG, Dhanarajan G, Nema S, Sen R, Roy U (2015) An antimicrobial metabolite from *Bacillus* sp.: significant activity against pathogenic bacteria including multidrug-resistant clinical strains. Front Microbiol 6:1335

Chapman SJ (1989) Oxidation of micronized elemental sulphur in soil. Plant Soil 116:69–76

Chartron J, Shiau C, Stout CD, Carroll KS (2007) 3′-Phosphoadenosine-5′-phosphosulfate reductase in complex with thioredoxin: a structural snapshot in the catalytic cycle. Biochemist 46:3942–3951

Chen KY, Morris JC (1972) Kinetics of oxidation of aqueous sulfide by oxygen. Environ Sci Technol 6:529–537

Chen TM, Kuschner WG, Gokhale J, Shofer S (2007) Outdoor air pollution: nitrogen dioxide, sulfur dioxide, and carbon monoxide health effects. Am J Med Sci 333:249–256

Chiang YL, Hsieh YC, Fang JY, Liu EH, Huang YC, Chuankhayan P, Chen CJ (2009) Crystal structure of adenylylsulfate reductase from *Desulfovibrio gigas* suggests a potential self-regulation mechanism involving the C terminus of the β-subunit. J Bacteriol 191:7597–7608

Ciferri O, Tiano P, Mastromei G (eds) (2000) Of microbes and art: the role of microbial communities in the degradation and protection of cultural heritage. Springer, New York

Collins MD, Weddel F (1986) Respiratory quinones of sulphate-reducing and sulphur-reducing bacteria: a systematic investigation. Syst Appl Microbiol 8:8–18

Constantin D, Bini A, Meletti E, Moldeus P, Monti D, Tomasi A (1996) Age-related differences in the metabolism of sulphite to sulphate and in the identification of sulphur trioxide radical in human polymorphonuclear leukocytes. Mech Ageing Dev 88:95–109

Cord-Ruwisch R, Kleinitz W, Widdel F (1987) Sulfate-reducing bacteria and their activities in oil production. J Pet Technol 39:97–106

Cord-Ruwisch R, Seitz HJ, Conrad R (1988) The capacity of hydrogenotrophic anaerobic bacteria to compete for traces of hydrogen depends on the redox potential of the terminal electron acceptor. Arch Microbiol 149:350–357

Costentin C (2008) Electrochemical approach to the mechanistic study of proton-coupled electron transfer. Chem Rev 108:2145–2179

Cypionka H (1987) Uptake of sulfate, sulfite and thiosulfate by proton–anion symport in *Desulfovibrio desulfuricans*. Arch Microbiol 148:144–149

Cypionka H (1994) Sulfate transport. In: Methods in enzymology, vol 243. Academic, San Diego, pp 3–14
Cypionka H (1995) Solute transport and cell energetics. In: Sulfate-reducing bacteria. Springer, Boston, pp 151–184
Czechowski MH, He SH, Nacro M, DerVartanian DV, Peck HD Jr, LeGall J (1984) A cytoplasmic nickel–iron hydrogenase with high specific activity from *Desulfovibrio multispirans* sp. N., a new species of sulfate reducing bacterium. Biochem Biophysical Res Comm 125:1025–1032
da Silva SM, Voordouw J, Leitao C, Martins M, Voordouw G, Pereira IA (2013) Function of formate dehydrogenases in *Desulfovibrio vulgaris* Hildenborough energy metabolism. Microbiology 159:1760–1769
Dahl C (2008) Microbial sulfur metabolism. In: Friedrich CG (ed). Springer, Berlin/New York, pp 151–169
Dahl C, Trüper HG (2001) Sulfite reductase and APS reductase from *Archaeoglobus fulgidus*. In: Methods in enzymology, vol 331. Academic, pp 427–441
Daskalakis MI, Magoulas A, Kotoulas G, Catsikis I, Bakolas A, Karageorgis AP, Rigas F (2013) *Pseudomonas* and *C. upriavidus* isolates induce calcium carbonate precipitation for biorestoration of ornamental stone. J Appl Microbiol 115:409–423
de Vito PC, Dreyfuss J (1964) Metabolic regulation of adenosine triphosphate sulfurylase in yeast. J Bacteriol 88:1341–1348
den Besten G, van Eunen K, Groen AK, Venema K, Reijngoud DJ, Bakker BM (2013) The role of short-chain fatty acids in the interplay between diet, gut microbiota, and host energy metabolism. J Lipid Res 54:2325–2340
Devalia JL, Rusznak C, Herdman MJ, Trigg CJ, Davies RJ, Tarraf H (1994) Effect of nitrogen dioxide and sulphur dioxide on airway response of mild asthmatic patients to allergen inhalation. Lancet 344:1668–1671
Devereux R, Delaney M, Widdel F, Stahl DA (1989) Natural relationships among sulfate-reducing eubacteria. J Bacteriol 171:6689–6695
Devereux R, Kane MD, Winfrey J, Stahl DA (1992) Genus- and group-specific hybridization probes for determinative and environmental studies of sulfate-reducing bacteria. Syst Appl Microbiol 15:601–609
Dhami NK, Reddy MS, Mukherjee A (2014) Application of calcifying bacteria for remediation of stones and cultural heritages. Front Microbiol 5:304
Dhillon A, Teske A, Dillon J, Stahl DA, Sogin ML (2003) Molecular characterization of sulfate-reducing bacteria in the Guaymas Basin. Appl Environ Microbiol 69:2765–2772
Didenko N, Skripnuk D (2014) The impact of energy resources on social development in Russia. In: Brebbia CA, Polonara F, Magaril ER, Passerini G (eds). Energy production and management in the 21st century: the quest for sustainable energy, vol 1. WIT Press, Southampton, pp 151–159
Dowling LM, Crewther WG, Inglis AS (1986) The primary structure of component 8c-1, a subunit protein of intermediate filaments in wool keratin. Relationships with proteins from other intermediate filaments. Biochem J 236:695–703
Duarte AG, Catarino T, White GF, Lousa D, Neukirchen S, Soares CM, Pereira IA (2018) An electrogenic redox loop in sulfate reduction reveals a likely widespread mechanism of energy conservation. Nat Commun 9:5448
Elshahed MS, Senko JM, Najar FZ, Kenton SM, Roe BA, Dewers TA, Krumholz LR (2003) Bacterial diversity and sulfur cycling in a mesophilic sulfide-rich spring. Appl Environ Microbiol 69:5609–5621
Enning D, Garrelfs J (2014) Corrosion of iron by sulfate-reducing bacteria: new views of an old problem. Appl Environ Microbiol 80:1226–1236
Fernandes P (2006) Applied microbiology and biotechnology in the conservation of stone cultural heritage materials. Appl Microbiol Biotechnol 73:291
Fewson CA (1988) Microbial metabolism of mandelate: a microcosm of diversity. FEMS Microbiol Rev 4:85–110

Fiala G, Stetter KO (1986) *Pyrococcus furiosus* sp. nov. represents a novel genus of marine heterotrophic archaebacteria growing optimally at 100 °C. Arch Microbiol 145:56–61

Fida TT, Voordouw J, Ataeian M, Kleiner M, Okpala GN, Mand J, Voordouw G (2018) Synergy of sodium nitroprusside and nitrate in inhibiting the activity of sulfate reducing bacteria in oil-containing bioreactors. Front Microbiol 9:981

Filipovic MR, Zivanovic J, Alvarez B, Banerjee R (2017) Chemical biology of H_2S signaling through persulfidation. Chem Rev 118:1253–1337

Finster K, Coates JD, Liesack W, Pfennig N (1997) *Desulfuromonas thiophila* sp. nov., a new obligately sulfur-reducing bacterium from anoxic freshwater sediment. Int J Syst Evol Microbiol 47:754–758

Fioletov VE, McLinden C, Krotkov N, Li C (2015) Lifetimes and emissions of SO_2 from point sources estimated from OMI. Geophys Res Lett 42:1969–1976

Fischer RX, Tillmanns E (1988) The equivalent isotropic displacement factor. Acta Crystallogr Sect C Cryst Struct Commun 44:775–776

Fitz RM, Cypionka H (1989) A study on electron transport-driven proton translocation in *Desulfovibrio desulfuricans*. Arch Microbiol 152:369–376

Forsberg CW (1980) Sulfide production from cysteine by *Desulfovibrio desulfuricans*. Appl Environ Microbiol 39:453–455

Foti M, Sorokin DY, Lomans B, Mussman M, Zacharova EE, Pimenov NV, Muyzer G (2007) Diversity, activity, and abundance of sulfate-reducing bacteria in saline and hypersaline soda lakes. Appl Environ Microbiol 73:2093–2100

Fröstl JM, Overmann J (2000) Phylogenetic affiliation of the bacteria that constitute phototrophic consortia. Arch Microbiol 174:50–58

Gahan J, Schmalenberger A (2014) The role of bacteria and mycorrhiza in plant sulfur supply. Front Plant Sci 5:723

Galushko AS, Schink B (2000) Oxidation of acetate through reactions of the citric acid cycle by *Geobacter sulfurreducens* in pure culture and in syntrophic coculture. Arch Microbiol 174:314–321

Gauri KL, Parks L, Jaynes J, Atlas R (1992) Removal of sulphated-crust from marble using sulphate-reducing bacteria. In: Stone cleaning and the nature, soiling and decay mechanisms of stone: proceedings of an international conference in Edinburgh, UK, 14–16 April 1992, pp 160–165

Gebhardt MR, Daniel TC, Schweizer EE, Allmaras RR (1985) Conservation tillage. Science 230:625–630

Ghosh W, Dam B (2009) Biochemistry and molecular biology of lithotrophic sulfur oxidation by taxonomically and ecologically diverse bacteria and archaea. FEMS Microbiol Rev 33:999–1043

Gigolashvili T, Kopriva S (2014) Transporters in plant sulfur metabolism. Front Plant Sci 5:442

Gilmour CC, Elias DA, Kucken AM, Brown SD, Palumbo AV, Schadt CW, Wall JD (2011) Sulfate-reducing bacterium *Desulfovibrio desulfuricans* ND132 as a model for understanding bacterial mercury methylation. Appl Environ Microbiol 77:3938–3951

Gitt MA, Wang LF, Doi RH (1985) A strong sequence homology exists between the major RNA polymerase sigma factors of *Bacillus subtilis* and *Escherichia coli*. J Biol Chem 260:7178–7185

Grein F, Ramos AR, Venceslau SS, Pereira IA (2013) Unifying concepts in anaerobic respiration: insights from dissimilatory sulfur metabolism. Biochimica et Biophysica Acta (BBA)-Bioenergetic 1827:145–160. doi: 10.1016/j.bbabio.2012.09.001

Guimaraes LHS (2012) Carbohydrates from biomass: sources and transformation by microbial enzymes, carbohydrates—comprehensive studies on glycobiology and glycotechnology. Chuan-Fa Chang, IntechOpen, pp 441–458. https://doi.org/10.5772/51576

Günal S, Hardman R, Kopriva S, Mueller JW (2019) Sulfation pathways from red to green. J Biol Chem 294:12293–12312

Gutknecht J, Walter A (1981) Transport of protons and hydrochloric acid through lipid bilayer membranes. Biochim Biophys Acta (BBA)-Biomemb 641:183–188

Haas KL, Franz KJ (2009) Application of metal coordination chemistry to explore and manipulate cell biology. Chem Rev 109(10):4921–4960

Hammes F, Verstraete W (2002) Key roles of pH and calcium metabolism in microbial carbonate precipitation. Rev Environ Sci Biotechnol 1:3–7

Hao X, Ma K (2003) Minimal sulfur requirement for growth and sulfur-dependent metabolism of the hyperthermophilic archaeon *Staphylothermus marinus*. Archaea:191–197

Hausmann B, Pelikan C, Herbold CW, Köstlbacher S, Albertsen M, Eichorst SA, Stingl U (2018) Peatland acidobacteria with a dissimilatory sulfur metabolism. ISME J 12:1729

Hazeu W, Batenburg-Van der Vegte WH, Bos P, Van der Pas RK, Kuenen JG (1988) The production and utilization of intermediary elemental sulfur during the oxidation of reduced sulfur compounds by *Thiobacillus ferrooxidans*. Arch Microbiol 150:574–579

Hébert A, Forquin-Gomez MP, Roux A, Aubert J, Junot C, Heilier JF, Beckerich JM (2013) New insights into sulfur metabolism in yeasts as revealed by studies of *Yarrowia lipolytica*. Appl Environ Microbiol 79:1200–1211

Hedderich R, Klimmek O, Kröger A, Dirmeier R, Keller M, Stetter KO (1998) Anaerobic respiration with elemental sulfur and with disulfides. FEMS Microbiol Rev 22:353–381

Heggendorn FL, Fraga AGM, Ferreira DDC, Gonçalves LS, Lione VDOF, Lutterbach MTS (2018) Sulfate-reducing bacteria: biofilm formation and corrosive activity in endodontic files. Int J Dentistry 2018:1–12

Heidelberg JF, Seshadri R, Haveman SA, Hemme CL, Paulsen IT, Kolonay JF, Daugherty SC (2004) The genome sequence of the anaerobic, sulfate-reducing bacterium *Desulfovibrio vulgaris* Hildenborough. Nat Biotechnol 22(5):554

Henrichs SM, Reeburgh WS (1987) Anaerobic mineralization of marine sediment organic matter: rates and the role of anaerobic processes in the oceanic carbon economy. Geomicrobiol J 5 (3–4):191–237

Herrmann J, Ravilious GE, McKinney SE, Westfall CS, Lee SG, Baraniecka P, Jez JM (2014) Structure and mechanism of soybean ATP sulfurylase and the committed step in plant sulfur assimilation. J Biol Chem 289(15):10919–10929

Heselmeyer K, Fischer U, Krumbein WE, Warscheid T (1991) Application of *Desulfovibrio vulgaris* for the bioconversion of rock gypsum crusts into calcite. Bioforum 1(2):89

Hocking WP, Stokke R, Roalkvam I, Steen IH (2014) Identification of key components in the energy metabolism of the hyperthermophilic sulfate-reducing archaeon *Archaeoglobus fulgidus* by transcriptome analyses. Front Microbiol 5:95

Hryniewicz M, Sirko A, Pałucha A, Böck A, Hulanicka D (1990) Sulfate and thiosulfate transport in *Escherichia coli* K-12: identification of a gene encoding a novel protein involved in thiosulfate binding. J Bacteriol 172(6):3358–3366

Huber SC, Huber JL, Campbell WH, Redinbaugh MG (1992) Apparent dependence of the light activation of nitrate reductase and sucrose-phosphate synthase activities in spinach leaves on protein synthesis. Plant Cell Physiol 33(5):639–646

Huber O, Korn R, McLaughlin J, Ohsugi M, Herrmann BG, Kemler R (1996) Nuclear localization of β-catenin by interaction with transcription factor LEF-1. Mech Dev 59(1):3–10

Huber R, Sacher M, Vollmann A, Huber H, Rose D (2000) Respiration of arsenate and selenate by hyperthermophilic archaea. Syst Appl Microbiol 23(3):305–314

Hügler M, Wirsen CO, Fuchs G, Taylor CD, Sievert SM (2005) Evidence for autotrophic CO_2 fixation via the reductive tricarboxylic acid cycle by members of the subdivision of proteobacteria. J Bacteriol 187(9):3020–3027

Jain RK, Kapur M, Labana S, Lal B, Sarma PM, Bhattacharya D, Thakur IS (2005) Microbial diversity: application of microorganisms for the biodegradation of xenobiotics. Curr Sci 89 (1):101–112

Jaramillo ML, Abanto M, Quispe RL, Calderón J, del Valle LJ, Talledo M, Ramírez P (2012) Cloning, expression and bioinformatics analysis of ATP sulfurylase from *Acidithiobacillus ferrooxidans* ATCC 23270 in *Escherichia coli*. Bioinformatics 8(15):695

Jeanjean R, Broda E (1977) Dependence of sulphate uptake by *Anacystis nidulans* on energy, on osmotic shock and on sulphate starvation. Arch Microbiol 114(1):19–23

Jeanthon C, L'Haridon S, Cueff V, Banta A, Reysenbach AL, Prieur D (2002) *Thermodesulfobacterium hydrogeniphilum* sp. nov., a thermophilic, chemolithoautotrophic, sulfate-reducing bacterium isolated from a deep-sea hydrothermal vent at Guaymas Basin, and emendation of the genus *Thermodesulfobacterium*. Int J Syst Evol Microbiol 52 (3):765–772

Jiménez-López C, Rodriguez-Navarro C, Piñar G, Carrillo-Rosúa FJ, Rodriguez-Gallego M, Gonzalez-Muñoz MT (2007) Consolidation of degraded ornamental porous limestone stone by calcium carbonate precipitation induced by the microbiota inhabiting the stone. Chemosphere 68(10):1929–1936

Jørgensen BB (1990) A thiosulfate shunt in the sulfur cycle of marine sediments. Science 249 (4965):152–154

Jørgensen BB, Bang M, Blackburn TH (1990) Anaerobic mineralization in marine sediments from the Baltic Sea–North Sea transition. Mar Ecol Prog Ser 59:39–54

Jørgensen BB, Findlay AJ, Pellerin A (2019) The biogeochemical sulfur cycle of marine sediments. Front Microbiol 10:849

Jormakka M, Törnrot S, Byrne B, Iwata S (2002) Molecular basis of proton motive force generation: structure of formate dehydrogenase-N. Science 295(5561):1863–1868

Joutey NT, Bahafid W, Sayel H, El Ghachtouli N (2013) Biodegradation: involved microorganisms and genetically engineered microorganisms. Biodegradation—Life of Science, pp 289–320. doi: 10.5772/56194

Jroundi F, Gómez-Suaga P, Jimenez-Lopez C, González-Muñoz MT, Fernandez-Vivas MA (2012) Stone-isolated carbonatogenic bacteria as inoculants in bioconsolidation treatments for historical limestone. Sci Total Environ 425:89–98

Junier P, Junier T, Podell S, Sims DR, Detter JC, Lykidis A, Bernier-Latmani R (2010) The genome of the Gram-positive metal- and sulfate-reducing bacterium *Desulfotomaculum reducens* strain MI-1. Environ Microbiol 12(10):2738–2754

Kety (1979) Disorders of the human brain. Sci Am 241(3):202–218

Kim BH, Kim HJ, Hyun MS, Park DH (1999) Direct electrode reaction of Fe(III)-reducing bacterium, *Shewanella putrefaciens*. J Microbiol Biotechnol 9:127–131

Klein L, Bruno R, Bavassano B, Rosenbauer H (1993) Anisotropy and minimum variance of magnetohydrodynamic fluctuations in the inner heliosphere. J Geophys Res Space Physics 98 (A10):17461–17466

Klenk R, Blieske U, Dieterle V, Ellmer K, Fiechter S, Hengel I, Lux-Steiner MC (1997) Properties of $CuInS_2$ thin films grown by a two-step process without H_2S. Solar Energy Mater Solar Cell 49(1–4):349–356

Kodama Y, Watanabe K (2003) Isolation and characterization of a sulfur-oxidizing chemolithotroph growing on crude oil under anaerobic conditions. Appl Environ Microbiol 69(1):107–112

Krämer M, Cypionka H (1989) Sulfate formation via ATP sulfurylase in thiosulfate- and sulfite-disproportionating bacteria. Arch Microbiol 151(3):232–237

Kreke B, Cypionka H (1992) Protonmotive force in freshwater sulfate-reducing bacteria, and its role in sulfate accumulation in *Desulfobulbus propionicus*. Arch Microbiol 158(3):183–187

Kuhad RC, Singh A (2013) Biotechnology for environmental management and resource recovery. Springer, New Delhi, pp 191–218

L'Haridon S, Cilia V, Messner P, Raguenes G, Gambacorta A, Sleytr UB, Jeanthon C (1998) *Desulfurobacterium thermolithotrophum* gen. nov., sp. nov., a novel autotrophic, sulphur-reducing bacterium isolated from a deep-sea hydrothermal vent. Int J Syst Evol Microbiol 48 (3):701–711

Laanbroek HJ, Stal LJ, Veldkamp H (1978) Utilization of hydrogen and formate by *Campylobacter* sp. under aerobic and anaerobic conditions. Arch Microbiol 119(1):99–102

Laanbroek HJ, Smit AJ, Nulend GK, Veldkamp H (1979) Competition for L-glutamate between specialised and versatile *Clostridium* sp. Arch Microbiol 120(1):61–66
Labrado AL, Brunner B, Bernasconi SM, Peckmann J (2019) Formation of large native sulfur deposits does not require molecular oxygen. Front Microbiol 10:24
Langer S, Hashimoto M, Hobl B, Mathes T, Mack M (2013) Flavoproteins are potential targets for the antibiotic roseoflavin in *Escherichia coli*. J Bacteriol 195(18):4037–4045
Larson LJ, Kuno M, Tao FM (2000) Hydrolysis of sulfur trioxide to form sulfuric acid in small water clusters. J Chem Phys 112(20):8830–8838
Laue H, Friedrich M, Ruff J, Cook AM (2001) Dissimilatory sulfite reductase (desulfoviridin) of the taurine-degrading, non-sulfate-reducing bacterium *Bilophila wadsworthia* RZATAU contains a fused DsrB–DsrD subunit. J Bacteriol 183(5):1727–1733
Leathen WW, Kinsel NA, Braley SA Sr (1956) *Ferrobacillus ferrooxidans*: a chemosynthetic autotrophic bacterium. J Bacteriol 72(5):700
Leavitt WD, Bradley AS, Santos AA, Pereira IAC, Johnston DT (2015) Sulfur isotope effects of dissimilatory sulfite reductase. Front Microbiol 6:1392
Leyva Salas M, Mounier J, Valence F, Coton M, Thierry A, Coton E (2017) Antifungal microbial agents for food biopreservation—a review. Microorganism 5(3):37
Liamleam W, Annachhatre AP (2007) Electron donors for biological sulfate reduction. Biotechnol Adv 25(5):452–463
Lie TJ, Godchaux W, Leadbetter ER (1999) Sulfonates as terminal electron acceptors for growth of sulfite-reducing bacteria (*Desulfitobacterium* spp.) and sulfate-reducing bacteria: effects of inhibitors of sulfidogenesis. Appl Environ Microbiol 65(10):4611–4617
Lodish H, Berk A, Kaiser CA, Krieger M, Scott MP, Bretscher A, Matsudaira P (2000) Molecular cell biology, 6th edn. W.H. Freeman and Company, New York
Lonergan DJ, Jenter HL, Coates JD, Phillips EJ, Schmidt TM, Lovley DR (1996) Phylogenetic analysis of dissimilatory Fe(III)-reducing bacteria. J Bacteriol 178(8):2402–2408
Long Y, Fang Y, Shen D, Feng H, Chen T (2016) Hydrogen sulfide (H_2S) emission control by aerobic sulfate reduction in landfill. Sci Rep 6:38103
Louis P, Flint HJ (2009) Diversity, metabolism and microbial ecology of butyrate-producing bacteria from the human large intestine. FEMS Microbiol Lett 294(1):1–8
Lovley DR, Phillips EJ, Lonergan DJ, Widman PK (1995) Fe(III) and S^0 reduction by *Pelobacter carbinolicus*. Appl Environ Microbiol 61(6):2132–2138
Macy JM, Schröder I, Thauer RK, Kröger A (1986) Growth the *Wolinella succinogenes* on H_2S plus fumarate and on formate plus sulfur as energy sources. Arch Microbiol 144(2):147–150
Majumder ELW, Wall JD (2017) Uranium bio-transformations: chemical or biological processes? Open J Inorg Chem 7(2):28–60
Makarova KS, Wolf YI, Mekhedov SL, Mirkin BG, Koonin EV (2005) Ancestral paralogs and pseudoparalogs and their role in the emergence of the eukaryotic cell. Nucleic Acids Res 33 (14):4626–4638
Manzella MP, Holmes DE, Rocheleau JM, Chung A, Reguera G, Kashefi K (2015) The complete genome sequence and emendation of the hyperthermophilic, obligate iron-reducing archaeon "*Geoglobus ahangari*" strain 234 T. Stand Genomic Sci 10(1):77
Martins M, Faleiro ML, Barros RJ, Veríssimo AR, Barreiros MA, Costa MC (2009) Characterization and activity studies of highly heavy metal resistant sulphate-reducing bacteria to be used in acid mine drainage decontamination. J Hazard Mater 166(2–3):706–713
Masuda S, Bao Z, Okubo T, Sasaki K, Ikeda S, Shinoda R, Minamisawa K (2016) Sulfur fertilization changes the community structure of rice root, and soil-associated bacteria. Microbes Environ 31(1):70–75
Matejuk A, Leng Q, Begum MD, Woodle MC, Scaria P, Chou ST, Mixson AJ (2010) Peptide-based antifungal therapies against emerging infections. Drugs Future 35(3):197
Matias PM, Pereira IA, Soares CM, Carrondo MA (2005) Sulphate respiration from hydrogen in *Desulfovibrio* bacteria: a structural biology overview. Prog Biophys Mol Biol 89(3):292–329

May E, Webster A, Inkpen R, Zamarreño D, Kuever J, Rudolph C, Ranalli G (2008) The BIOBRUSH project for bioremediation of Heritage stone. In: Heritage microbiology and science: microbes, monuments and maritime materials. Royal Society of Chemistry, Cambridge, pp 76–93

McNamara CJ, Mitchell R (2005) Microbial deterioration of historic stone. Front Ecol Environ 3 (8):445–451

Md Zain WS, Salleh H, Insyirah, N., Abdullah, A. (2018). Natural biocides for mitigation of sulphate reducing bacteria. Int J Corrosion 2018: 1–7

Melim Kristen M, Shinglman Penelope J, Boston Diana E, Northup Michael N, Spilde J, Michael Queen L (2001) Evidence for microbial involvement in pool finger precipitation, Hidden Cave, New Mexico. Geomicrobiol J 18(3):311–329

Menendez JA, Menendez EM, Iglesias MJ, Garcıa A, Pis JJ (1999) Modification of the surface chemistry of active carbons by means of microwave-induced treatments. Carbon 37 (7):1115–1121

Meyer B, Kuever J (2008) Homology modeling of dissimilatory APS reductases (AprBA) of sulfur-oxidizing and sulfate-reducing prokaryotes. PLoS One 3(1):e1514

Miralles-Robledillo JM, Torregrosa-Crespo J, Martínez-Espinosa RM, Pire C (2019) DMSO reductase family: phylogenetics and applications of extremophiles. Int J Mol Sci 20(13):3349

Módis K, Bos EM, Calzia E, Van Goor H, Coletta C, Papapetropoulos A, Szabo C (2014) Regulation of mitochondrial bioenergetic function by hydrogen sulfide. Part II. Pathophysiological and therapeutic aspects. Br J Pharmacol 171(8):2123–2146

Möller-Zinkhan D, Börner G, Thauer RK (1989) Function of methanofuran, tetrahydromethanopterin, and coenzyme F 420 in *Archaeoglobus fulgidus*. Arch Microbiol 152(4):362–368

Müller A, Krebs B (eds) (2016) Sulfur: its significance for chemistry, for the geo-, bio-, and cosmosphere and technology, vol 5. Elsevier, Burlington

Muñoz-Dorado J, Marcos-Torres FJ, García-Bravo E, Moraleda-Muñoz A, Pérez J (2016) Myxobacteria: moving, killing, feeding, and surviving together. Front Microbiol 7:781

Mussmann M, Richter M, Lombardot T, Meyerdierks A, Kuever J, Kube M, Amann R (2005) Clustered genes related to sulfate respiration in uncultured prokaryotes support the theory of their concomitant horizontal transfer. J Bacteriol 187(20):7126–7137

Myers CR, Nealson KH (1988a) Bacterial manganese reduction and growth with manganese oxide as the sole electron acceptor. Science 240(4857):1319–1321

Myers CR, Nealson KH (1988b) Microbial reduction of manganese oxides: interactions with iron and sulfur. Geochim Cosmochim Acta 52(11):2727–2732

Nanninga HJ, Gottschal JC (1987) Properties of *Desulfovibrio carbinolicus* sp. nov. and other sulfate-reducing bacteria isolated from an anaerobic-purification plant. Appl Environ Microbiol 53(4):802–809

Nazina TN, Rozanova EP, Kuznetsov SI (1985) Microbial oil transformation processes accompanied by methane and hydrogen-sulfide formation. Geomicrobiol J 4(2):103–130

Newsome L, Morris K, Lloyd JR (2014) The biogeochemistry and bioremediation of uranium and other priority radionuclides. Chem Geol 363:164–184

Nguyen JL, Schwartz J, Dockery DW (2014) The relationship between indoor and outdoor temperature, apparent temperature, relative humidity, and absolute humidity. Indoor Air 24 (1):103–112

Noreña CZ, Rigon RT (2018) Effect of blanching on enzyme activity and bioactive compounds of blackberry. Braz Arch Biol Technol 61:e18180018. https://doi.org/10.1590/1678-4324-2018180018

Northup E, Kathleen H, Lavoie D (2001) Geomicrobiology of caves: a review. Geomicrobiol J 18 (3):199–222

Norvaišienė R, Miniotaitė R, Stankevičius V (2003) Climatic and air pollution effects on building facades. Mater Sci (Medžiagotyra) 16(1):80–85

Novotný C, Kapralek F (1979) Participation of quinone and cytochrome b in tetrathionate reductase respiratory chain of *Citrobacter freundii*. Biochem J 178(1):237–240
O'Brien JM, Wolkin RH, Moench TT, Morgan JB, Zeikus JG (1984) Association of hydrogen metabolism with unitrophic or mixotrophic growth of *Methanosarcina barkeri* on carbon monoxide. J Bacteriol 158(1):373–375
O'Flaherty V, Lens P, Leahy B, Colleran E (1998) Long-term competition between sulphate-reducing and methane-producing bacteria during full-scale anaerobic treatment of citric acid production wastewater. Water Res 32(3):815–825
Odom JM, Singleton R (1993) The sulfate-reducing bacteria: contemporary perspectives. Springer, New York, pp 189–210
Okafor N (2011) Taxonomy, physiology, and ecology of aquatic microorganisms. In: Environmental microbiology of aquatic and waste systems. Springer, Dordrecht, pp 47–107
Oliveira TF, Vonrhein C, Matias PM, Venceslau SS, Pereira IA, Archer M (2008) The crystal structure of *Desulfovibrio vulgaris* dissimilatory sulfite reductase bound to DsrC provides novel insights into the mechanism of sulfate respiration. J Biol Chem 283(49):34141–34149
Oliveira TF, Franklin E, Afonso JP, Khan AR, Oldham NJ, Pereira IA, Archer M (2011) Structural insights into dissimilatory sulfite reductases: structure of desulforubidin from *Desulfomicrobium norvegicum*. Front Microbiol 2(71):71
Oliver MJ (2014) Why we need GMO crops in agriculture. Missouri Med 111(6):492
Ollivier B, Zeyen N, Gales G, Hickman-Lewis K, Gaboyer F, Benzerara K, Westall F (2018) Importance of prokaryotes in the functioning and evolution of the present and past geosphere and biosphere. In: Prokaryotes and evolution. Springer, Cham, pp 57–129
Oltmann LF, Stouthamer AH (1975) Reduction of tetrathionate, trithionate and thiosulphate, and oxidation of sulphide in *Proteus mirabilis*. Arch Microbiol 105(1):135–142
Oltmann LF, Van der Beek EG, Stouthamer AH (1975) Reduction of inorganic sulphur compounds by facultatively aerobic bacteria. Plant Soil 43(1–3):153–169
Ortega-Calvo JJ, Ariño X, Hernandez-Marine M, Saiz-Jimenez C (1995) Factors affecting the weathering and colonization of monuments by phototrophic microorganisms. Sci Total Environ 167(1–3):329–341
Ouattara AS, Traore AS, Garcia JL (1992) Characterization of *Anaerovibrio burkinabensis* sp. nov., a lactate fermenting bacterium isolated from rice field soils. Int J Syst Evol Microbiol 42 (3):390–397
Oyaizu H, Woese CR (1985) Phylogenetic relationships among the sulfate respiring bacteria, myxobacteria and purple bacteria. Syst Appl Microbiol 6(3):257–263
Ozawa K, Tsapin AI, Nealson KH, Cusanovich MA, Akutsu H (2000) Expression of a tetraheme protein, *Desulfovibrio vulgaris* Miyazaki F cytochrome c3, in *Shewanella oneidensis* MR-1. Appl Environ Microbiol 66(9):4168–4171
Pace NR (1997) A molecular view of microbial diversity and the biosphere. Science 276 (5313):734–740
Palková Z (2004) Multicellular microorganisms: laboratory versus nature. EMBO Rep 5 (5):470–476
Parey K, Fritz G, Ermler U, Kroneck PM (2013) Conserving energy with sulfates around 100 C—structure and mechanism of key metal enzymes in hyperthermophilic *Archaeoglobus fulgidus*. Metallomics 5(4):302–317
Parkes RJ, Gibson GR, Mueller-Harvey I, Buckingham WJ, Herbert RA (1989) Determination of the substrates for sulphate-reducing bacteria within marine and esturaine sediments with different rates of sulphate reduction. Microbiology 135(1):175–187
Peck HD (1961) Enzymatic basis for assimilatory and dissimilatory sulfate reduction. J Bacteriol 82 (6):933–939
Peck HD Jr (1960) Adenosine 5′-phosphosulfate as an intermediate in the oxidation of thiosulfate by *Thiobacillus thioparus*. Proc Natl Acad Sci U S A 46(8):1053
Peck HD, Odom M (1981) Anaerobic fermentations of cellulose to methane. In: Trends in the biology of fermentations for fuels and chemicals. Springer, Boston, pp 375–395

Pereira Roders A, Van Oers R (2011) Initiating cultural heritage research to increase Europe's competitiveness. J Cultural Herit Manag Sustain Dev 1(2):84–95

Pereira IA, Ramos AR, Grein F, Marques MC, Da Silva SM, Venceslau SS (2011a) A comparative genomic analysis of energy metabolism in sulfate reducing bacteria and archaea. Front Microbiol 2:69

Pereira PHF, Voorwald HCJ, Cioffi MOH, Mullinari DR, Da Luz SM, Da Silva MLCP (2011b) Sugarcane bagasse pulping and bleaching: thermal and chemical characterization. Biol Res 6 (3):2471–2482

Pester M, Knorr KH, Friedrich MW, Wagner M, Loy A (2012) Sulfate-reducing microorganisms in wetlands—fameless actors in carbon cycling and climate change. Front Microbiol 3:72

Pfennig N, Biebl H (1976) *Desulfuromonas acetoxidans* gen. nov. and sp. nov., a new anaerobic, sulfur-reducing, acetate-oxidizing bacterium. Arch Microbiol 110(1):3–12

Pfennig N, Widdel F (1981) Ecology and physiology of some anaerobic bacteria from the microbial sulfur cycle. In: Biology of inorganic nitrogen and sulfur. Springer, Berlin/Heidelberg, pp 169–177

Pihl TD, Black LK, Schulman BA, Maier RJ (1992) Hydrogen-oxidizing electron transport components in the hyperthermophilic archaebacterium *Pyrodictium brockii*. J Bacteriol 174 (1):137–143

Pikuta E, Lysenko A, Suzina N, Osipov G, Kuznetsov B, Tourova T, Laurinavichius K (2000) *Desulfotomaculum alkaliphilum* sp. nov., a new alkaliphilic, moderately thermophilic, sulfate-reducing bacterium. Int J Syst Evol Microbiol 50(1):25–33

Plugge CM, Zhang W, Scholten J, Stams AJ (2011) Metabolic flexibility of sulfate-reducing bacteria. Front Microbiol 2:81

Poyatos F, Morales F, Nicholson AW, Giordano A (2018) Physiology of biodeterioration on canvas paintings. J Cell Physiol 233(4):2741–2751

Prange A, Arzberger I, Engemann C, Modrow H, Schumann O, Trüper HG, Hormes J (1999) In situ analysis of sulfur in the sulfur globules of phototrophic sulfur bacteria by x-ray absorption near edge spectroscopy. Biochim Biophys Acta (BBA)-Gen Subj 1428(2–3):446–454

Prioretti L, Gontero B, Hell R, Giordano M (2014) Diversity and regulation of ATP sulfurylase in photosynthetic organisms. Front Plant Sci 5:597

Probst GS, Bousquet WF, Miya TS (1977) Correlation of hepatic metallothionein concentrations with acute cadmium toxicity in the mouse. Toxicol Appl Pharmacol 39(1):61–69

Rabus R, Hansen TA, Widdel F (2013) Dissimilatory sulfate- and sulfur-reducing prokaryotes. In: Rosenberg E, DeLong EF, Lory S, Stackebrandt E, Thompson F (eds) The prokaryotes. Springer, Berlin/Heidelberg, pp 309–404

Ramesh MV, Sirakova TD, Kolattukudy PE (1995) Cloning and characterization of the cDNAs and genes (mep20) encoding homologous metalloproteinases from *Aspergillus flavus* and *A. fumigatus*. Gene 165(1):121–125

Ranalli G, Chiavarini M, Guidetti V, Marsala F, Matteini M, Zanardini E, Sorlini C (1997) The use of microorganisms for the removal of sulphates on artistic stoneworks. Int Biodeterior Biodegrad 40(2–4):255–261

Ranalli G, Matteini M, Tosini I, Zanardini E, Sorlini C (2000) Bioremediation of cultural heritage: removal of sulphates, nitrates and organic substances. In: Of microbe art. Springer, Boston, pp 231–245

Ranalli G, Alfano G, Belli C, Lustrato G, Colombini MP, Bonaduce I, Zanardini E, Abbruscato P, Cappitelli F, Sorlini C (2005) Biotechnology applied to cultural heritage: biorestoration of frescoes using viable bacterial cells and enzymes. J Appl Microbiol 98:73–83

Rapp-Giles BJ, Casalot L, English RS, Ringbauer JA, Dolla A, Wall JD (2000) Cytochrome c3 mutants of *Desulfovibrio desulfuricans*. Appl Environ Microbiol 66(2):671–677

Reece SY, Hodgkiss JM, Stubbe J, Nocera DG (2006) Proton-coupled electron transfer: the mechanistic underpinning for radical transport and catalysis in biology. Philos Trans R Soc B Biol Sci 361(1472):1351–1364

Roden EE, Lovley DR (1993) Dissimilatory Fe(III) reduction by the marine microorganism *Desulfuromonas acetoxidans*. Appl Environ Microbiol 59(3):734–742

Rodriguez-Navarro C, Rodriguez-Gallego M, Chekroun KB, Gonzalez-Munoz MT (2003) Conservation of ornamental stone by *Myxococcus xanthus* induced carbonate biomineralization. Appl Environ Microbiol 69(4):2182–2193

Rogerio-Candelera MA, Lazzari M, Cano E (eds) (2013) Science and technology for the conservation of cultural heritage. CRC, Boca Raton

Rosnes JT, Torsvik T, Lien T (1991) Spore-forming thermophilic sulfate-reducing bacteria isolated from North Sea oil field waters. Appl Environ Microbiol 57(8):2302–2307

Roy AB, Trudinger PA (1970) The chemistry of some sulphur compounds. The biochemistry of inorganic compounds of sulphur. Cambridge University Press, Cambridge, pp 7–29

Saiz-Jimenez C (1993) Deposition of airborne organic pollutants on historic buildings. Atmos Environ Part B Urban Atmos 27(1):77–85

Sáiz-Jiménez C, Garcia-Rowe J, Rodriguez-Hidalgo JM (1991) Biodeterioration of polychrome Roman mosaics. Int Biodeterior 28(1–4):65–79

Sanmartín P, Bosch-Roig P (2019) Biocleaning to remove graffiti: a real possibility? Advances towards a complete protocol of action. Coating 9(2):104

Santoro C, Zarkout K, Le Hô AS, Mirambet F, Gourier D, Binet L, Griesmar P (2014) New highlights on degradation process of verdigris from easel paintings. Appl Phys A 114 (3):637–645

Santos AA, Venceslau SS, Grein F, Leavitt WD, Dahl C, Johnston DT, Pereira IA (2015) A protein trisulfide couples dissimilatory sulfate reduction to energy conservation. Science 350 (6267):1541–1545

Schäfer H, Mathey D, Hugo F, Bhakdi S (1986) Deposition of the terminal C5b-9 complement complex in infarcted areas of human myocardium. J Immunol 137(6):1945–1949

Schauder R, Widdel F, Fuchs G (1987) Carbon assimilation pathways in sulfate-reducing bacteria II. Enzymes of a reductive citric acid cycle in the autotrophic *Desulfobacter hydrogenophilus*. Arch Microbiol 148(3):218–225

Schauder R, Preuß A, Jetten M, Fuchs G (1988) Oxidative and reductive acetyl CoA/carbon monoxide dehydrogenase pathway in *Desulfobacterium autotrophicum*. Arch Microbiol 151 (1):84–89

Scheffers DJ, Pinho MG (2005) Bacterial cell wall synthesis: new insights from localization studies. Microbiol Mol Biol Rev 69(4):585–607

Schiffer A, Parey K, Warkentin E, Diederichs K, Huber H, Stetter KO, Ermler U (2008) Structure of the dissimilatory sulfite reductase from the hyperthermophilic archaeon *Archaeoglobus fulgidus*. J Mol Biol 379(5):1063–1074

Schillinger U, Geisen R, Holzapfel WH (1996) Potential of antagonistic microorganisms and bacteriocins for the biological preservation of foods. Trends Food Sci Technol 7(5):158–164

Schlesner H, Lawson PA, Collins MD, Weiss N, Wehmeyer U, Völker H, Thomm M (2001) *Filobacillus milensis* gen. nov., sp. nov., a new halophilic spore-forming bacterium with Orn-D-Glu-type peptidoglycan. Int J Syst Evol Microbiol 51(2):425–431

Schmidt A, Jäger K (1992) Open questions about sulfur metabolism in plants. Annu Rev Plant Biol 43(1):325–349

Schmitz ML, Baeuerle PA (1991) The p65 subunit is responsible for the strong transcription activating potential of NF-kappa B. EMBO J 10(12):3805–3817

Scholten JC, Stams AJ (2000) Isolation and characterization of acetate-utilizing anaerobes from freshwater sediment. Microb Ecol 40(4):292–299

Segerer AH, Stetter KO, Klink F (1985) Two contrary modes of chemolithotrophy in the same archaebacterium. Nature 313(6005):787

Segerer AH, Neuner A, Kristjansson JK, Stetter KO (1986) *Acidianus infernus* gen. nov., sp. nov., and *Acidianus brierleyi* comb. nov.: facultatively aerobic, extremely acidophilic thermophilic sulfur-metabolizing archaebacteria. Int J Syst Evol Microbiol 36(4):559–564

Segerer AH, Trincone A, Gahrtz M, Stetter KO (1991) *Stygiolobus azoricus* gen. nov., sp. nov. represents a novel genus of anaerobic, extremely thermoacidophilic archaebacteria of the order *Sulfolobales*. Int J Syst Evol Microbiol 41(4):495–501

Selig M, Schönheit P (1994) Oxidation of organic compounds to CO_2 with sulfur or thiosulfate as electron acceptor in the anaerobic hyperthermophilic archaea *Thermoproteus tenax* and *Pyrobaculum islandicum* proceeds via the citric acid cycle. Arch Microbiol 162(4):286–294

Selig M, Xavier KB, Santos H, Schönheit P (1997) Comparative analysis of Embden–Meyerhof and Entner–Doudoroff glycolytic pathways in hyperthermophilic archaea and the bacterium *Thermotoga*. Arch Microbiol 167(4):217–232

Shen Y, Buick R (2004) The antiquity of microbial sulfate reduction. Earth Sci Rev 64 (3–4):243–272

Siebers B, Hensel R (1993) Glucose catabolism of the hyperthermophilic archaeum *Thermoproteus tenax*. FEMS Microbiol Lett 111(1):1–7

Silva M, Rosado T, Teixeira D, Candeias A, Caldeira AT (2015) Production of green biocides for cultural heritage. Novel Biotechnol Solution. Int J Conserv Sci 6:519–530

Silva-Castro GA, Uad I, Gonzalez-Martinez A, Rivadeneyra A, Gonzalez-Lopez J, Rivadeneyra MA (2015) Bioprecipitation of calcium carbonate crystals by bacteria isolated from saline environments grown in culture media amended with seawater and real brine. Biomed Res Int 2015:1–12

Simon J, Kroneck PM (2013) Microbial sulfite respiration. In: Advances in microbial physiology, vol 62. Academic, New York, pp 45–117

Sirko A, Hryniewicz M, Hulanicka D, Böck A (1990) Sulfate and thiosulfate transport in *Escherichia coli* K-12: nucleotide sequence and expression of the cys TWAM gene cluster. J Bacteriol 172(6):3351–3357

Skyring GW (1987) Sulfate reduction in coastal ecosystems. Geomicrobiol J 5(3–4):295–374

Smutná T, Pilarová K, Tarábek J, Tachezy J, Hrdý I (2014) Novel functions of an iron–sulfur flavoprotein from *Trichomonas vaginalis* hydrogenosomes. Antimicrob Chemo 58 (6):3224–3232

Soffritti I, D'Accolti M, Lanzoni L, Volta A, Bisi M, Mazzacane S, Caselli E (2019) The potential use of microorganisms as restorative agents: an update. Sustainability 11(14):3853

Speich N, Dahl C, Heisig P, Klein A, Lottspeich F, Stetter KO, Trüper HG (1994) Adenylylsulphate reductase from the sulphate-reducing archaeon *Archaeoglobus fulgidus*: cloning and characterization of the genes and comparison of the enzyme with other iron–sulphur flavoproteins. Microbiology 140(6):1273–1284

Sperling D, Kappler U, Wynen A, Dahl C, Trüper HG (1998) Dissimilatory ATP sulfurylase from the hyperthermophilic sulfate reducer *Archaeoglobus fulgidus* belongs to the group of homo-oligomeric ATP sulfurylases. FEMS Microbiol Lett 162(2):257–264

Stahlmann R, Korte M, Van Loveren H, Vos JG, Thiel R, Neubert D (1992) Abnormal thymus development and impaired function of the immune system in rats after prenatal exposure to aciclovir. Arch Toxicol 66(8):551–559

Steinsbu BO, Thorseth IH, Nakagawa S, Inagaki F, Lever MA, Engelen B, Pedersen RB (2010) *Archaeoglobus sulfaticallidus* sp. nov., a thermophilic and facultatively lithoautotrophic sulfate-reducer isolated from black rust exposed to hot ridge flank crustal fluids. Int J Syst Evol Microbiol 60(12):2745–2752

Sterflinger K, Piñar G (2013) Microbial deterioration of cultural heritage and works of art—tilting at windmills? Appl Microbiol Biotechnol 97(22):9637–9646

Stetter KO, Huber R, Blöchl E, Kurr M, Eden RD, Fielder M, Vance I (1993) Hyperthermophilic archaea are thriving in deep North Sea and Alaskan oil reservoirs. Nature 365(6448):743

Sutyak KE, Wirawan RE, Aroutcheva AA, Chikindas ML (2008) Isolation of the *Bacillus subtilis* antimicrobial peptide subtilosin from the dairy product–derived *Bacillus amyloliquefaciens*. J Appl Microbiol 104(4):1067–1074

Syed (2006) Atmospheric corrosion of materials. EMI J Eng Res 11(1):1–24

Szabo C, Ransy C, Módis K, Andriamihaja M, Murghes B, Coletta C, Bouillaud F (2014) Regulation of mitochondrial bioenergetic function by hydrogen sulfide. Part I. Biochemical and physiological mechanisms. Br J Pharmacol 171(8):2099–2122
Taylor J, Parkes RJ (1983) The cellular fatty acids of the sulphate-reducing bacteria, *Desulfobacter* sp., *Desulfobulbus* sp. and *Desulfovibrio desulfuricans*. Microbiology 129(11):3303–3309
Thom J, Anderson GM (2008) The role of thermochemical sulfate reduction in the origin of Mississippi Valley–type deposits. I. Experimental results. Geofluids 8(1):16–26
Tian H, Gao P, Chen Z, Li Y, Li Y, Wang Y, Ma T (2017) Compositions and abundances of sulfate-reducing and sulfur-oxidizing microorganisms in water-flooded petroleum reservoirs with different temperatures in China. Front Microbiol 8:143
Tilman D, Cassman KG, Matson PA, Naylor R, Polasky S (2002) Agricultural sustainability and intensive production practices. Nature 418(6898):671
Toran L, Harris RF (1989) Interpretation of sulfur and oxygen isotopes in biological and abiological sulfide oxidation. Geochim Cosmochim Acta 53(9):2341–2348
Ueki A, Suto T (1979) Cellular fatty acid composition of sulfate-reducing bacteria. J Gen Appl Microbiol 25(3):185–196
Ullrich TC, Blaesse M, Huber R (2001) Crystal structure of ATP sulfurylase from *Saccharomyces cerevisiae*, a key enzyme in sulfate activation. EMBO J 20(3):316–329
Urbina J, Patil A, Fujishima K, Paulino-Lima IG, Saltikov C, Rothschild LJ (2019) A new approach to biomining: bioengineering surfaces for metal recovery from aqueous solutions. Sci Rep 9 (1):1–11
Urzì C (1999) On microbes and art: the role of microbial communities in the degradation and protection of cultural heritage. A report on the International Conference on Microbiology and Conservation (ICMC 1999). Environ Microbiol 1(6):551–553
Valente FM, Saraiva LM, LeGall J, Xavier AV, Teixeira M, Pereira IA (2001) A membrane-bound cytochrome c3: a type II cytochrome c3 from *Desulfovibrio vulgaris* Hildenborough. Chembiochem 2(12):895–905
Van Driessche AE, Stawski TM, Benning LG, Kellermeier M (2017) Calcium sulfate precipitation throughout its phase diagram. In: New perspectives on mineral nucleation and growth. Springer, Cham, pp 227–256
Velho RV, Medina LFC, Segalin J, Brandelli A (2011) Production of lipopeptides among *Bacillus* strains showing growth inhibition of phytopathogenic fungi. Folia Microbiol 56(4):297
Velikova V, Yordanov I, Edreva A (2000) Oxidative stress and some antioxidant systems in acid rain–treated bean plants: protective role of exogenous polyamines. Plant Sci 151(1):59–66
Venceslau SS, Lino RR, Pereira IA (2010) The Qrc membrane complex, related to the alternative complex III, is a menaquinone reductase involved in sulfate respiration. J Biol Chem 285 (30):22774–22783
Venceslau SS, Stockdreher Y, Dahl C, Pereira IAC (2014) The "bacterial heterodisulfide" DsrC is a key protein in dissimilatory sulfur metabolism. Biochim Biophysica Acta (BBA)-Bioenerg 1837 (7):1148–1164
Venkateswaran K, Chung S, Allton J, Kern R (2004) Evaluation of various cleaning methods to remove *Bacillus* spores from spacecraft hardware materials. Astrobiology 4(3):377–390
Vieille C, Zeikus GJ (2001) Hyperthermophilic enzymes: sources, uses, and molecular mechanisms for thermostability. Microbiol Mol Biol Rev 65(1):1–43
Vimr ER, Kalivoda KA, Deszo EL, Steenbergen SM (2004) Diversity of microbial sialic acid metabolism. Microbiol Mol Biol Rev 68(1):132–153
Vladár P, Rusznyák A, Márialigeti K, Borsodi AK (2008) Diversity of sulfate-reducing bacteria inhabiting the rhizosphere of *Phragmites australis* in Lake Velencei (Hungary) revealed by a combined cultivation-based and molecular approach. Microb Ecol 56(1):64–75
Vorholt JA, Hafenbradl D, Stetter KO, Thauer RK (1997) Pathways of autotrophic CO_2 fixation and of dissimilatory nitrate reduction to N_2O in *Ferroglobus placidus*. Arch Microbiol 167(1):19–23
Wang G, Zhang R, Gomez ME, Yang L, Zamora ML, Hu M, Li J (2016) Persistent sulfate formation from London fog to Chinese haze. Proc Natl Acad Sci 113(48):13630–13635

Warscheid T, Braams J (2000) Biodeterioration of stone: a review. Int Biodeterior Biodegrad 46 (4):343–368

Warthmann R, Cypionka H (1990) Sulfate transport in *Desulfobulbus propionicus* and *Desulfococcus multivorans*. Arch Microbiol 154(2):144–149

Warthmann R, Van Lith Y, Vasconcelos C, McKenzie JA, Karpoff AM (2000) Bacterially induced dolomite precipitation in anoxic culture experiments. Geology 28(12):1091–1094

Widdel F, Bak F (1992) Gram-negative mesophilic sulfate-reducing bacteria. In: The prokaryotes. Springer, New York, pp 3352–3378

Widdel F, Rabus R (2001) Anaerobic biodegradation of saturated and aromatic hydrocarbons. Curr Opin Biotechnol 12(3):259–276

Wilmes P, Bond PL (2004) The application of two-dimensional polyacrylamide gel electrophoresis and downstream analyses to a mixed community of prokaryotic microorganisms. Environ Microbiol 6(9):911–920

Wilson LG, Asahi T, Bandurski RS (1961) Yeast sulfate-reducing system I. Reduction of sulfate to sulfite. J Biol Chem 236(6):1822–1829

Woese CR, Fox GE (1977) Phylogenetic structure of the prokaryotic domain: the primary kingdoms. Proc Natl Acad Sci 74(11):5088–5090

Woese CR, Magrum LJ, Fox GE (1978) Archaebacteria. J Mol Evol 11(3):245–252

Wolf J, Stark H, Fafenrot K, Albersmeier A, Pham TK, Müller KB, Kouril T (2016) A systems biology approach reveals major metabolic changes in the thermoacidophilic archaeon *Sulfolobus solfataricus* in response to the carbon source L-fucose versus D-glucose. Mol Microbiol 102(5):882–908

Yant WP, Schrenk HH, Patty FA (1936) A plant study of urine sulfate determinations as a measure of benzene exposure. J Ind Hyg Toxicol 18:349–356

Ye J, Zhang R, Nielsen S, Joseph SD, Huang D, Thomas T (2016) A combination of biochar–mineral complexes and compost improves soil bacterial processes, soil quality, and plant properties. Front Microbiol 7:372. https://doi.org/10.3389/fmicb.2016.00372

Yoon S, Sanford RA, Löffler FE (2013) *Shewanella* spp. use acetate as an electron donor for denitrification but not ferric iron or fumarate reduction. Appl Environ Microbiol 79 (8):2818–2822

Zellner G, Messner P, Winter J, Stackebrandt E (1998) *Methanoculleus palmolei* sp. nov., an irregularly coccoid methanogen from an anaerobic digester treating wastewater of a palm oil plant in North-Sumatra, Indonesia. Int J Syst Evol Microbiol 48(4):1111–1117

Zhang Y, Wang X, Zhen Y, Mi T, He H, Yu Z (2017) Microbial diversity and community structure of sulfate-reducing and sulfur-oxidizing bacteria in sediment cores from the East China Sea. Front Microbiol 8:2133

Zhu T, Dittrich M (2016) Carbonate precipitation through microbial activities in natural environment, and their potential in biotechnology: a review. Front Bioeng Biotechnol 4:4

Zillig W, Stetter KO, Prangishvilli D, Schäfer W, Wunderl S, Janekovic D, Palm P (1982) Desulfurococcaceae, the second family of the extremely thermophilic, anaerobic, sulfur-respiring Thermoproteales. Zentralblatt Für Bakteriologie Mikrobiologie Und Hygiene: I. Abt. Originale C: Allgemeine, Angewandte Und Ökologische Mikrobiologie 3(2):304–317

Zinder SH, Brock TD (1978) Dimethyl sulphoxide reduction by micro-organisms. Microbiology 105(2):335–342

Zöphel A, Kennedy MC, Beinert H, Kroneck PMH (1988) Investigations on microbial sulfur respiration. Arch Microbiol 150(1):72–77

Microbiological Tools for Cultural Heritage Conservation

8

Amrita Kumari Panda, Rojita Mishra, and Satpal Singh Bisht

Abstract

Culture and cultural heritage are the imprints of human civilization and architectural depiction of society and cultures. Antique documents and cultural heritages such as historic buildings, monuments, manuscripts and paintings are brittle and undergo physical, chemical and biological deterioration during the course of time. Escalating air pollution and global warming are the main cause of deterioration of stone monuments and artworks. The deterioration process can be restored by employing various microbiological tools such as biocleaning, biomineralization, biocementation and biofilm formation. This chapter summarizes eco-friendly microbiological approaches used to restore cultural heritages, archaeological sites and wall paintings.

Keywords

Cultural heritage · Biocleaning · Biomineralization · Biocementation · Deterioration

The original version of this chapter was revised. A correction to this chapter can be found at https://doi.org/10.1007/978-981-15-3401-0_11

A. K. Panda (✉)
Department of Biotechnology, Sant Gahira Guru University, Ambikapur, Chhattisgarh, India

R. Mishra
Department of Botany, Polasara Science College, Ganjam, Odisha, India

S. S. Bisht
Department of Zoology, Kumaun University, Nainital, Uttarakhand, India

A. N. Yadav et al. (eds.), *Microbial Biotechnology Approaches to Monuments of Cultural Heritage*, https://doi.org/10.1007/978-981-15-3401-0_8

8.1 Introduction

Biodeterioration plays significant role in the degradation of stone in monuments, historic buildings and archaeological sites (Sterflinger 2010; Saiz-Jimenez 1999). Several cultural heritage materials are at menace of biodeterioration including metals (Berk et al. 2001), ceramics (Sand and Bock 1991), glass (Schabereiter-Gurtner et al. 2001), paintings (Rubio and Bolivar 1997), wood (Bjordal et al. 1999), synthetic polymers (Gu et al. 1998) and mummified bodies (Arya et al. 2001). Microorganisms that have been established as the causative agents in weakening of stone include bacteria (Cuzman et al. 2010), archaea (Rolleke et al. 1998), cyanobacteria (Crispim and Gaylarde 2005), algae (Tomaselli et al. 2000), fungi (Adeyemi and Gadd 2005; Burford et al. 2003) and lichens (Adamo and Violante 2000; Seneviratne and Indrasena 2006). There are also reports that stone objects may support novel communities of microorganisms (e.g. alkaliphiles, halophiles and endoliths) that function in the biodeterioration process (Saiz-Jimenez and Laiz 2000).

Microorganisms add considerably to the overall deterioration phenomena on stone, concrete, mortar, slurries and paint coatings, glass and metals used in architecture (Pinar and Sterflinger 2009).

The environment especially temperate humid climate influences the colonization of microbes in a particular habitat. These monuments get ruined and discoloured by the growth and metabolic activity of microbes, forming a biofilm over the surface, thus lessening its aesthetic value (Fig. 8.1). The slow and unalterable deterioration of cultural heritages speeds up during the last century due to climate change and increasing air pollution (Jroundi et al. 2017). Many attempts have been tried to slow down the deterioration of heritage sites with both organic and inorganic products. Microbiological tools proved as a novel approach to hinder these irreversible damages by inducing biomineralization and biofilm formation. The present chapter is an attempt to summarize various microbiological tools used to conserve the historic monuments, cultural heritages and archaeological sites (Fig. 8.1).

8.2 Microorganisms Involved in Biodeterioration

There is a popular saying that microbes have found their way into museums, caves and ruins: 'Almost any cultural artefact is prone to colonization by microorganisms'. Stone cultural heritage materials are at threat of biodeterioration caused by varied populations of microorganisms existing in biofilms. The microbial metabolites of these biofilms are responsible for the deterioration of the underlying substratum and may lead to physical weakening and discolouration of stone. Air pollutants in urban environments accelerate biodeterioration by serving as an additional nutrient source for the microorganisms.

Biodeterioration can be defined as *any undesirable change in the properties of a material caused by the vital activities of living organisms* (Hueck 1965). Various environmental factors, viz. high temperatures, heavy rainfall and high relative humidity levels, favour the growth of a wide variety of microbial communities on various ancient monuments, rock paints, etc. (Biswas et al. 2013). Further, fungi are the most accepted harmful microbes among all the microbial communities. It has been comprehended that the rock paints in caves are the most valuable heritages which are suffering from serious fungal attacks in the past few years.

Fig. 8.1 (a) Wall paintings of Ajanta Caves, Maharashtra, with heavy fungal growth. (b) Sita Devi Temple, Deorbija, Chhattisgarh, India, made of sandstone biodeteriorated due to thick biofilm of fungi and lichen. (c) Photograph of Adhinath Jain Temple, Pavagadh, showing visible microbial growth

8.2.1 Fungi Associated with Biodeterioration

Fungi play more hazardous role in the biodeterioration of stone monuments because of their complex metabolic activities and capability to grow on range of substrates and tolerate extremes of environmental conditions, establishing mutualistic association with cyanobacteria and algae. Many fungal species such as *Alternaria*, *Acremonium*, *Aspergillus*, *Aureobasidium*, *Arthobotrys*, *Cladosporium*, *Drechslera*, *Curvularia*, *Fusarium*, *Mucor*, *Rhizopus*, *Helminthosporium*, *Trichothecium*, *Penicillium* and *Trichoderma* have been reported as common fungi involved in biological decay of stone monuments (Farooq et al. 2015). Fungi cause harm to the monumental stones and artworks in several ways. They produce many organic acids such as oxalic, citric, gluconic and lactic acids during their metabolic activities. These acids function as chelating agent and augment the solubility of metals and forms complexes with the mineral cations present on the stone surfaces (Gómez-Alarcón and Munoz 1995; Dakal and Cameotra 2012). Some fungi weaken the stone texture by invasion of fungal hyphae and alternate contraction and expansion of the thallus under variable environmental humidity conditions resulting in mechanical deterioration of monuments (Gadd 2004).

8.2.2 Bacteria Associated with Biodeterioration

Biocorrosion of stone can occur through the action of cyanobacteria (photoautotrophs) and nitrifying and sulphur-oxidizing bacteria (chemolithoautotrophs). Nitrous and nitric acid excreted by nitrifying bacteria (*Nitrosomonas* spp. and *Nitrobacter* spp.) and sulphuric acid excreted by sulphur-oxidizing bacteria (*Thiobacillus* spp.) lead to stone dissolution and salt formation. These acids react with stone constituents and produce sulphate-based crusts which get precipitated within the pores of the stone and upon recrystallization exert substantial stress in the pore walls (Fernandes 2006). Bacterial species like *Geodermatophilaceae* strains and *Micromonospora* strains were isolated from monumental stones (Urzì et al. 2001). There are reports that *Pseudomonas fluorescens* was found in the caves of Lascaux in association with *Fusarium* mainly due to illumination and visitor's breath (Allemand and Bahn 2005).

8.3 Microbial Metabolism and Its Effect on Cultural Heritage

Microbes and microbial metabolism play major role in conservation and deterioration of cultural heritage. Sulphate reduction is an important microbial metabolism that can be used for removing black crusts on stone and artworks. Many researchers used this as a tool for conservation of stone works (Bosch-Roig and Ranalli 2014; Alfano et al. 2011; Polo et al. 2010; Cappitelli et al. 2007; Ranalli et al. 2000; Gauri et al. 1992). Archaeological objects can be saved by using the iron reduction mechanism. Some bacterial strains like *Desulfitobacterium hafniense* and *Geobacter sulfurreducens* have the capacity to convert rusted iron layer into stable corrosion product by biogenic mineral precipitation approach (Comensoli et al. 2017; Cote et al. 2015; Junier and Joseph 2017). Oxalogenesis is the process of formation of metal oxalates by fungi, and this can be used for the restoration of archaeological objects (Joseph et al. 2012a, b). Microbially induced calcium carbonate precipitation is used in limestone monument repairs. Many researchers used this for filling of pores and cracks in monuments (Jonkers 2011; Dhami et al. 2013; Cheng et al. 2013). Microbial metabolism has been evolved as technology that restores many cultural heritages with the help of bacteria and fungi (Fig. 8.2).

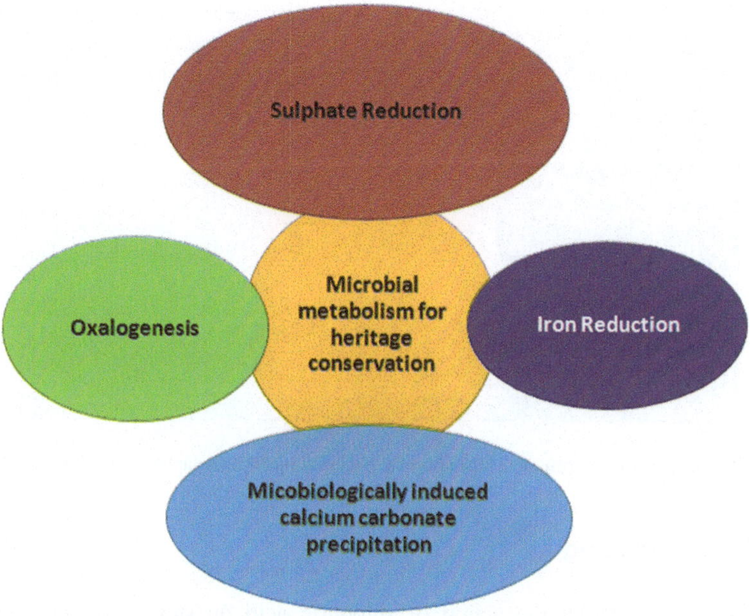

Fig. 8.2 Metabolism of microorganisms applied for conservation of heritage

8.4 Biofilms and Its Role in Heritage Conservation

Biofilm is a well-organized microbial mat with cells entrapped in a matrix formed by extracellular polymer (O'Toole and Ghannoum 2004). Biofilm matrix forms a transport barrier, which may prevent the penetration of corrosive agents (such as oxygen, chloride, etc.) and decrease their contact with the metal surface, thus reducing corrosion (Fig. 8.3). The possible mechanisms may involve (1) removal of oxygen by bacterial respiration, (2) antimicrobials secreted by growth-inhibiting corrosion-causing bacteria within biofilms [e.g. sulphate-reducing bacteria (SRB) corrosion inhibition by gramicidin S-producing *Bacillus brevis* biofilm] and (3) production of shielding layer by biofilms (e.g. *Bacillus licheniformis* biofilm produces on aluminium surface a sticky protective layer of γ-polyglutamate) (Zuo 2007) (Fig. 8.3).

8.5 Tools for Conservation

Microbial metabolism and biogeochemical cycling together can develop many technologies that can be used as tool for the conservation of cultural heritage. Microorganisms can be used for the synthesis of inorganic components in an eco-friendly approach. Various tools of conservation are described below.

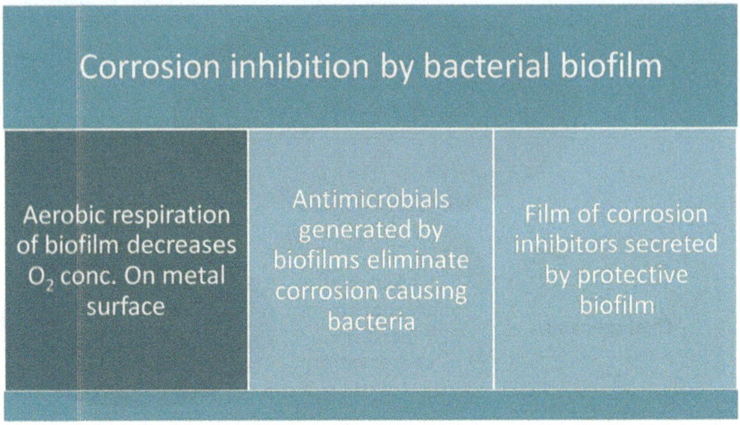

Fig. 8.3 Mechanism of corrosion inhibition by bacterial biofilm (Adapted from Zuo 2007)

8.5.1 Biocleaning

Gypsum is formed by the action of sulphuric acid on stonework. Gypsum, in association with airborne pollutants, causes the blackening of sheltered areas (Ausset et al. 2000; Bugini et al. 2000). Chemical and mechanical treatment methods are used to remove black crust. These methods are not satisfactory so far as conservation is taken into account. Microorganisms are used to remove these black crusts. This process of conservation of monuments by using microorganisms is called as biocleaning. Sulphate-reducing bacteria are the group of microorganisms mainly involved to restore the sulphate-based crusts with the help of their metabolism. This biocleaning work started from 1989; since then many researchers worked in this field, and various biocleaning techniques are developed by many researchers for conservation of cultural heritage. The working principle and efficacy of biocleaning is described in a simplified manner in Fig. 8.4 (Bosch-Roig and Ranalli 2014) (Table 8.1).

8.5.2 Safety Aspects of Biocleaning

Cleaning of cultural heritage by microorganisms raised some questions regarding the safety aspect of this advance cleaning technologies. It is more important to evaluate the safety aspect of various biocleaning technologies. More research must be conducted to ensure that these technologies are really safe and soft on our cultural heritage. Short-term and long-term surveillance and monitoring on any development will provide significant information regarding safety aspect of these technologies. Proper guidelines and information about the safety aspect of biocleaning provide a new platform for the development and commercialization of these technologies and

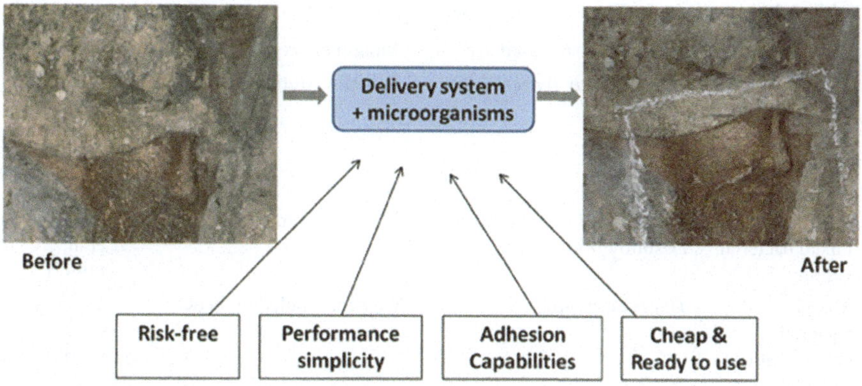

Fig. 8.4 Main characteristics and effects of biocleaning process (bacteria and enzyme) on a fragment on Conversione di S. Efisio e battaglia, fresco (fourteenth century), at Pisa Camposanto Monumentale (Italy). Courtesy of Bosch-Roig P and Ranalli G (2014)

development of eco-friendly and sustainable restoration strategy for conservation of cultural heritage (Bosch-Roig and Ranalli 2014).

8.5.3 Biomineralization

Biomineralization is the process of formation of mineral as a result of oxidation or reduction of a metal species. Microbial activity releases variety of mineral precipitates such as oxalates, carbonates and oxides that act as cementing layer and enhance the durability of stone monuments (Gadd and Dyer 2017). Stone monuments specially made up of limestone and marble can be conserved by bacterially induced calcite precipitation (Jroundi et al. 2017). Fungal biomineralization is mainly due to the production of low molecular weight organic acids, i.e. acetic, citric, gluconic, fumaric, glyoxylic and oxalic acids. These organic acids are present as their conjugated bases at environmental pH and form precipitation with different metals (Bindschedler et al. 2016) (Table 8.2).

8.5.4 Biorestoration

There are evidences that fungal biofilms can be a potent bioprotector, and when these layers are removed, they can accelerate the deterioration of stone monuments and cultural heritage (McIlroy et al. 2012; Gadd 2017). There are reports that endolithic lichen *Bagliettoa baldensis* plays an important role in the protection of carboniferous limestone surfaces from rainfall-induced weathering (McIlroy de la Rosa et al. 2014). The organic biofilm derived from various endolithic lichens acts as a waterproof layer and is mainly responsible for the well-preserved state of certain ancient heritage buildings. Concha-Lozano et al. 2012 reported that colonization of

Table 8.1 List of biocleaning technology used for conservation of monuments

Bioformulation	Microorganisms used	Monument conserved	References
Whole bacterial cell/ protease/ collagenase (1:3:10)	*Pseudomonas stutzeri* A29 strain	Spinello Aretino fresco, Italy	Ranalli et al. (2005)
Pure and mixed bacterial culture	*Desulfovibrio* desulfuricans Desulfovibrio *vulgaris*	Marble column and statue	Ranalli et al. (1997)
Mixed bacterial culture with suitable nutrient	*Thiobacillus* sp.	Concrete fouled by lichen	De Graef et al. (2005)
Hydrobiogel-97 and Carbogel	*Desulfovibrio vulgaris* subsp. *vulgaris* ATCC 29579	EniTecnologie, San Donato Milanese, Italy, and CST, Vicenza, Italy	Cappitelli et al. (2006)
Carbogel	*Desulfovibrio vulgaris* subsp. *vulgaris* ATCC 29579	Milan Cathedral (Italy)	Cappitelli et al. (2007)
Carbogel	*Pseudomonas pseudoalcaligenes* KF707 strain and *Desulfovibrio vulgaris* ATCC 29579 cells	Matera Cathedral	Alfano et al. (2011)
Viable cells	*Pseudomonas stutzeri*	Spinello Aretino from Pisa's cemetery	Antonioli et al. (2005)
Viable cells	*Pseudomonas stutzeri*, A29	Camposanto Monumentale in Pisa, Italy	Lustrato et al. (2012)
Carbogel	*Desulfovibrio vulgaris* ATCC 29579 cells	English cemetery in Florence	Gioventù et al. (2012)
Advanced agar-gauze biogel system	*Pseudomonas stutzeri* A29	The Vatican Museums, Cristo che salva Pietro dalle acque – La Navicella, Pisa Cathedral Cupola, Incarnato Pisa Cathedral Cupola, Incarnato	Ranalli et al. (2019)

Verrucaria nigrescens and *Caloplaca aurantia* can fill the pores in the limestone with a dense network of lichenized fungal hyphae which waterproof the stone and could act as a sulphate contamination barrier. Slavík et al. 2017 reported that biologically initiated rock crust (BIRC) formed by fungi, lichens and green algae was less erodible and has 3–35 times higher tensile strength and 2–33 times lower water absorption than the subsurface sandstone.

Biopatinas inhibits the corrosion process and provides a very high-level protection to archaeological and artistic metal artefacts. The use of microorganisms for protection of stone monuments is novel approach studied extensively during the last

Table 8.2 Potential microorganisms that induce biomineralization

Metabolic pathway	Microorganism	Mechanism of bioprecipitation	References
Heterotrophic pathway	*Desulfovibrio* sp.	Sulphate reduction releases sulphide and metabolic CO_2 as a result of which calcium carbonate precipitation occurs	Braissant et al. (2007)
	Myxococcus xanthus	Oxidative deamination of amino acids produces ammonia and CO_2 that create alkaline microenvironment which favours calcium carbonate precipitation	Ettenauer et al. (2011); Jroundi et al. (2010); Rodriguez-Navarro et al. (2003)
	Pseudomonas aeruginosa *Diaphorobacter nitroreducens*	Denitrification causes elevated medium pH that facilitates the biosynthesis of calcium carbonate	Erşan et al. (2015)
	Bacillus licheniformis *Bacillus megaterium*	Urea hydrolysis increases the pH of the medium by producing ammonia. Availability of a calcium source in the surrounding medium leads to precipitation of calcium carbonate	Helmi et al. (2016)
Autotrophic pathway	*Synechococcus* PCC8806	Removal of carbon dioxide during oxygenic photosynthesis from bicarbonate solutions results in carbonate production	Zhu et al. (2015)

decade in contrast to the currently used organic protective coatings, which simply create a barrier against aggressive environments in a non-selective way. Copper-oxalate patina from *Beauveria bassiana* (naturally occurring fungus) was used for bioprotection of a copper artefact (Joseph et al. 2012). Microorganisms such as sulphate-reducing bacteria can be used for the removal of black crusts on stone artworks. *Desulfovibrio vulgaris* subsp. *vulgaris* ATCC 29579 is a sulphate-reducing bacterium used successfully for the removal of black crusts from the altered Candoglia marble from the cathedral of Milan (Cappitelli et al. 2006).

8.6 Projects Using Novel Microbiological Tools for Cultural Heritage Conservation

Conservation and protection of cultural heritage is an important aspect of research in modern era. European research has dedicated 20 years in the conservation of cultural heritage. Many projects were sanctioned under this mission named as Environment and climate programme. It is divided into six phases. During 1994–1998 a project named ROCK ART started which includes studies on environment, geology, geochemistry and microbiology aspects for conservation of prehistoric rock art. In the second phase, MICROCOR was started which is based on novel molecular tools for analysis of microbial communities grown on mural paintings for its conservation and restoration practices. Two more projects were funded in the second phase entitled NEW SURFACE and DETERIORATION in the developmental aspects of heritage conservation programme. During the Environment and sustainable development programme in phase 5 (1999–2002), many projects started and successfully completed like BIOBRUSH which includes bioremediation for building restoration of urban stone heritage, BIODAM which includes inhibitors on biofilm damage on mineral materials, BIOREINFORCE which is a biologically influenced calcite precipitation for monument stone reinforcement, CATS which mainly deals with cyanobacteria and associated microorganisms on Roman hypogen monuments, VIRDIO which mainly deals with determining the conditions to prevent weathering by microorganisms on ancient stained glass windows with protective glazing and GRAFFITAGE which was sanctioned in sixth phase (2003–2007) which includes development of a new anti-graffiti system for preventing damage to architecture of heritage material (Source: European Commission's Research DG, http://ec.europa.eu/research).

8.7 Conclusion and Future Prospects

This chapter explains that bioprotection of cultural heritages through microbiological tools is highly effective in comparison to physical and chemical restoration methods. Microbial biofilms and microbially induced biomineralization are novel and eco-friendly approaches for the conservation of cultural heritage. The interdisciplinary research among microbiologists and heritage conservators can facilitate global appliance of this new conservation methodology for preservation of many historic heritage sites. It is a cost-effective and eco-friendly approach where deterioration is reversed using microbes. Future conservation strategies mainly based on microbiological tools can definitely provide better conservation to monuments and heritages in an environment-friendly manner, and it also has lesser effect on climate change.

References

Adamo P, Violante P (2000) Weathering of rocks and neogenesis of minerals associated with lichen activity. Appl Clay Sci 16:229–256

Adeyemi AO, Gadd GM (2005) Fungal degradation of calcium-, lead- and silicon-bearing minerals. Biometals 18:269–281

Alfano G, Lustrato G, Belli C, Zanardini E, Cappitelli F, Mello E et al (2011) The bioremoval of nitrate and sulfate alterations on artistic stonework: the case-study of Matera cathedral after six years from the treatment. Int Biodeterior Biodegradation 65:1004–1011

Allemand L, Bahn PG (2005) Best way to protect rock art is to leave it alone. Nature 433:800

Antonioli P, Zapparoli G, Abbruscato P, Sorlini C, Ranalli G, Righetti PG (2005) Art-loving bugs: the resurrection of Spinello Aretino from Pisa's cemetery. Proteomics 5:2453–2459
Arya AA, Shah R, Sadasivan S (2001) Curr Sci 81:793–799
Ausset P, Lefèvre RA, Del Monte M (2000) Early mechanisms of development of sulfated black crusts on carbonate stone, pp 329–337. In: Fassina V (ed) Proceedings of the 9th international congress on deterioration and conservation of stone, Venice, Italy. Elsevier Science, Amsterdam
Berk SG, Mitchell R, Bobbie RJ, Nickels JS, White DC (2001) Int Biodeterior Biodegrad 48:167–175
Bindschedler S, Cailleau G, Verrecchia E (2016) Role of fungi in the biomineralization of calcite. Fortschr Mineral 6:41
Biswas J, Sharma K, Harris KK, Rajput Y (2013) Biodeterioration agents: bacterial and fungal diversity dwelling in or on the pre-historic rock-paints of Kabra-pahad, India. Iran J Microbiol 5:309
Bjordal CG, Nilsson T, Daniel G (1999) Int Biodeterior Biodegradation 43:63–71
Bosch-Roig P, Ranalli G (2014) The safety of biocleaning technologies for cultural heritage. Front Microbiol 5:155
Braissant O, Decho AW, Dupraz C, Glunk C, Przekop KM, Visscher PT (2007) Exopolymeric substances of sulfate-reducing bacteria: interactions with calcium at alkaline pH and implication for formation of carbonate minerals. Geobiology 5:401–411
Bugini R, Laurenzi Tabasso M, Realini M (2000) Rate of formation of black crusts on marble. A case study. J Cult Herit 1:111–116
Burford EP, Fomina M, Gadd GM (2003) Fungal involvement in bioweathering and biotransformation of rocks and minerals. Mineral Mag 67:1127–1155
Cappitelli F, Zanardini E, Ranalli G, Mello E, Daffonchio D, Sorlini C (2006) Improved methodology for bioremoval of black crusts on historical stone artworks by use of sulfate-reducing bacteria. Appl Environ Microbiol 72:3733–3737
Cappitelli F, Toniolo L, Sansonetti A, Gulotta D, Ranalli G, Zanardini E, Sorlini C (2007) Advantages of using microbial technology over traditional chemical technology in removal of black crusts from stone surfaces of historical monuments. Appl Environ Microbiol 73:5671–5675
Cheng L, Cord-Ruwisch R, Shahin MA (2013) Cementation of sand soil by microbially induced calcite precipitation at various degrees of saturation. Can Geotech J 50:81–90
Comensoli L et al (2017) Use of bacteria to stabilize archaeological iron. Appl Environ Microbiol 83:e03478–e03416
Concha-Lozano N, Gaudon P, Pages J, de Billerbeck G, Lafon D, Eterradossi O (2012) Protective effect of endolithic fungal hyphae on oolitic limestone buildings. J Cult Herit 13:120–127
Cote C, Rosas O, Basseguy R (2015) *Geobacter sulfurreducens*: an iron reducing bacterium that can protect carbon steel against corrosion? Corros Sci 94:104–113
Crispim CA, Gaylarde CC (2005) Cyanobacteria and biodeterioration of cultural heritage: a review. Microb Ecol 49:1–9
Cuzman OA, Ventura S, Sili C, Mascalchi C, Turchetti T, D'Acqui LP, Tiano P (2010) Biodiversity of phototrophic biofilms dwelling on monumental fountains. Microb Ecol 60:81–95
De Graef B, De Windt W, Dick J, Verstraete W, De Belie N (2005) Cleaning of concrete fouled by lichens with the aid of *Thiobacilli*. Mater Struct 38:875–882
McIlroy de la Rosa JP, Warke PA, Smith BJ (2012) Lichen-induced biomodification of calcareous surfaces: bioprotection versus biodeterioration. Prog Phys Geogr 37:325–351
McIlroy de la Rosa JP, Warke PA, Smith BJ (2014) The effects of lichen cover upon the rate of solutional weathering of limestone. Geomorphology 220:81–92
Dakal TC, Cameotra SS (2012) Microbially induced deterioration of architectural heritages: routes and mechanisms involved. Environ Sci Eur 24(1):36
Dhami NK, Reddy MS, Mukherjee A (2013) Biomineralization of calcium carbonates and their engineered applications: a review. Front Microbiol 4:314

Erşan YÇ, Belie N, Boon N (2015) Microbially induced CaCO3 precipitation through denitrification: an optimization study in minimal nutrient environment. Biochem Eng J 101:108–118

Ettenauer J, Piñar G, Sterflinger K, Gonzalez-Muñoz MT, Jroundi F (2011) Molecular monitoring of the microbial dynamics occurring on historical limestone buildings during and after the in situ application of different bio-consolidation treatments. Sci Total Environ 409:5337–5352

Farooq M, Hassan M, Gull F (2015) Mycobial deterioration of stone monuments of Dharmarajika, Taxila. J Microbiol Exp 2:36

Fernandes P (2006) Applied microbiology and biotechnology in the conservation of stone cultural heritage materials. Appl Microbiol Biotechnol 73:291

Gadd GM (2004) Mycotransformation of organic and inorganic substrates. Mycologist 18:60–70

Gadd GM (2017) Geomicrobiology of the built environment. Nat Microbiol 2:16275

Gadd GM, Dyer TD (2017) Bioprotection of the built environment and cultural heritage. Microb Biotechnol 10:1152–1156

Gauri KL, Parks L, Jaynes J, Atlas R (1992) Removal of sulfated-crusts from marble using sulphate-reducing bacteria. In: Webster RGM (ed) Proceedings of the international conference on stone cleaning and the nature, soiling and decay mechanisms of stone. Donheadd, Edinburgh, pp 160–165

Gioventù E, Lorenzi P, Improta MC, Cappitelli F (2012) Bacterial cleaning technology for marble surfaces affected by black crust: comparison with chemical and laser treatments. 12th International Congress on the Deterioration and Conservation of Stone, Columbia University, New York

Gómez-Alarcón G, Munoz ML (1995) Flores M excretion of organic acids by fungal strains isolated from decayed sandstone. Int Biodeterior Biodegrad 34:169–180

Gu JD, Mitton DB, Ford TE, Mitchell R (1998) Biodegradation 9:39–45

Helmi FM, Elmitwalli HR, Elnagdy SM, El-Hagrassy AF (2016) Calcium carbonate precipitation induced by ureolytic bacteria *Bacillus licheniformis*. Ecol Eng 90:367–371

Hueck-van der Plas EH (1965) The biodeterioration of materials as a part of hylobiology. Mater Org 1:5–34

Jonkers H (2011) Bacteria-based self-healing concrete. Heron 56:1–12

Joseph E, Cario S, Simon A, Wörle M, Mazzeo R, Junier P, Job D (2012) Protection of metal artifacts with the formation of metal–oxalates complexes by *Beauveria bassiana*. Front Microbiol 2:270

Joseph E et al (2012a) Protection of metal artefacts with the formation of metal-oxalates complexes by *Beauveria bassiana*. Front Microbiol 2:270

Joseph E, Simon A, Mazzeo R, Job D, W€orle M (2012b) Spectroscopic characterization of an innovative biological treatment for corroded metal artefacts. J Raman Spectrosc 43:1612–1616

Jroundi F, Fernández-Vivas A, Rodriguez-Navarro C, Bedmar EJ, González-Muñoz MT (2010) Bioconservation of deteriorated monumental calcarenite stone and identification of bacteria with carbonatogenic activity. Microb Ecol 60:39–54

Jroundi F, Schiro M, Ruiz-Agudo E, Elert K, Martín-Sánchez I, González-Muñoz MT, Rodriguez-Navarro C (2017) Protection and consolidation of stone heritage by self-inoculation with indigenous carbonatogenic bacterial communities. Nat Commun 8:279

Junier P, Joseph E (2017) Microbial biotechnology approaches to mitigating the deterioration of construction and heritage materials. Microb Biotechnol 10:1145–1148

Lustrato G, Alfano G, Andreotti A, Colombini MP, Ranalli G (2012) Fast biocleaning of mediaeval frescoes using viable bacterial cells. Int Biodeterior Biodegrad 69:51–61

O'Toole GA, Ghannoum MA (2004) Introduction to biofilms: conceptual themes. In Microbial Biofilms. American Society of Microbiology, pp 1–3

Pinar G, Sterflinger K (2009) Microbes and building materials. In: Cornejo DN, Haro JL (eds) Building materials: properties, performance and applications. Nova Science Publishers, New York, pp 163–188

Polo A, Cappitelli F, Brusetti L, Villa F, Giacomucci L, Ranalli G, Sorlini C (2010) Feasibility of removing surface deposits on stone using biological and chemical remediation methods. Microb Ecol 60:1–14

Ranalli G, Chiavarini M, Guidetti E, Marsala F, Matteini M, Zanardini E, Sorlini C (1997) The use of microorganism for the removal of sulphates on artistic stone works. Int Biodeterior Biodegrad 40:255–261

Ranalli G, Matteini M, Tosini I, Zanardini E, Sorlini C (2000) Bioremediation of cultural heritage: removal of sulphates, nitrates and organic substances. In: Ciferri O, Tiano P, Mastromei G (eds) Of microbes and art. Springer, Boston

Ranalli G, Alfano G, Belli C, Lustrato G, Colombini MP, Bonaduce I, Zanardini E, Abbruscato P, Cappitelli F, Sorlini C (2005) Biotechnology applied to cultural heritage: biorestoration of frescoes using viable bacterial cells and enzymes. J Appl Microbiol 98:73–83

Ranalli G et al (2019) Onsite advanced biocleaning system for historical wall paintings using new agar-gauze bacteria gel. J Appl Microbiol 126:1785–1796

Ranalli G, Zanardini E, Rampazzi L, Corti C, Andreotti A, Colombini MP, Bosch-Roig P, Lustrato G, Giantomassi C, Zari D, Virilli P (2019) Onsite advanced biocleaning system for historical wall paintings using new agar-gauze bacteriagel. J Appl Microbiol 126(6):1785–1796

Rodriguez-Navarro C, Rodriguez-Gallego M, Chekroun KB, Gonzalez Muñoz MT (2003) Conservation of ornamental stone by *Myxococcus xanthus*-induced carbonate biomineralization. Appl Environ Microbiol 69:2182–2193

Rolleke S, Witte A, Wanner G, Lubitz W (1998) Medieval wall painting-an habitat for Archaea: identification of Archaea by denaturing gradient gel electrophoresis (DGGE) of PCR-amplified gene fragments coding 16S rRNA in a medieval painting. Int Biodeterior Biodegr 41:85–92

Rubio RF, Bolivar FC (1997) Int Biodeterior Biodegradation 40:161–169

Saiz-Jimenez C (1999) Geomicrobiol J 16:27–37

Saiz-Jimenez C, Laiz L (2000) Int Biodeterior Biodegradation 46:319–326

Sand W, Bock E (1991) Int Biodeterior Biodegradation 27:175–183

Schabereiter-Gurtner C, Pinar G, Lubitz W, Rolleke S (2001) J Microbiol Methods 47:345–354

Seneviratne G, Indrasena IK (2006) Nitrogen fixation in lichens is important for improved rock weathering. J Biosci 31:639–643

Slavík M, Bruthans J, Filippi M, Schweigstillová J, Falteisek L, Jaroslav Řihošek J (2017) Biologically-initiated rock crust on sandstone: mechanical and hydraulic properties and resistance to erosion. Geomorphology 278:298–313

Sterflinger K (2010) Fungi: their role in deterioration of cultural heritage. Fungal Biol Rev 24:47–55

Tomaselli L, Lamenti G, Bosco M, Tiano P (2000) Biodiversity of photosynthetic micro-organisms dwelling on stone monuments. Int Biodeterior Biodegr 46(3):251–258

Urzì C, Brusetti L, Salamone P, Sorlini C, Stachebrandt E, Doffonchio D (2001) Biodiversity of Geodermatophilaceae isolated from altered stones and monuments in the Mediterranean basin. Environ Microbiol 3:471–479

Zhu T, Paulo C, Merroun ML, Dittrich M (2015) Potential application of biomineralization by *Synechococcus* PCC8806 for concrete restoration. Ecol Eng 82:459–468

Zuo R (2007) Biofilms: strategies for metal corrosion inhibition employing microorganisms. Appl Microbiol Biotechnol 76:1245–1253

9 Biotechnology to Restoration and Conservation

Lamha Kumar, Neha Kapoor, and Archana Tiwari

Abstract

Biotechnology is a broad area of biology that involves the technological application of living cells, organisms and systems to make or develop products. It exploits cellular and bimolecular processes to innovate technologies and products that improve the standard of living and the planet's health. Biotechnology has been essential in the betterment of medicine, industry and agriculture. It has great potential for the conservation and preservation of cultural monuments. With a rapid increase in population and pollution, cultural monuments are now being degraded and eroded away at a rapid pace. Some have blackened, some have lost their shine and some are left ruined by acid rain. Mechanical methods such as washing have not proven to be a success in the conservation of these monuments. Recent work using biotechnology has proven to be a better alternative in the conservation and preservation of monuments. Biotechnology research in restoration work of cultural monuments develops in two directions, one which focuses on the development of accurate diagnostic techniques for the identification and characterization of bio-deteriogens and alterations and the other which focuses on the development of innovative restoration methods by the employment of new products. Bacteria that reduce sulphur were used for black crust of marble from the Cathedral of Florence. The *Desulfovibrio desulfuricans* bacteria have been used for the removal of black patina which contains large amounts of sulphates. Denitrifying bacteria, for example, *Pseudomonas stutzeri*, have been used for the removal of nitrate pollutants.

L. Kumar · N. Kapoor
Hindu College, University of Delhi, Delhi, India

A. Tiwari (✉)
Amity University, Noida, Uttar Pradesh, India
e-mail: atiwari9@amity.edu

A. N. Yadav et al. (eds.), *Microbial Biotechnology Approaches to Monuments of Cultural Heritage*, https://doi.org/10.1007/978-981-15-3401-0_9

Keywords

Conservation · Cultural monuments · Deteriogens · Microbiomes · Restoration

9.1 Introduction

A monument is a large structure that has been erected to commemorate a notable event or person in history. According to the Archaeological Survey of India (ASI), any rock sculpture, cave, monolith, monument, erection, inscription or structure which has existed for at least hundred years and is of artistic, historical or archaeological interest is referred to as an ancient monument. Monuments serve as a reflection of history and help understand and observe the changes that have occurred in society. Monuments help realize the different habits and traditions of people of different eras and the reasons that led to the development of cities, societies and different traditions. Further, they reflect conflicts, wars, the prosperity of society and the economic conditions of the time that they were built. They serve a very important role in promoting tourism and appreciation of art and culture of different timelines of the world. Due to the current rate at which population and therefore pollution are increasing, all aspects of the environment are being affected, including the historical building. Monuments are being severely affected in many ways.

The monuments made of marble are turning black or brownish yellow due to pollution. Pollutants like dust, SPM (suspended particulate matter) and other species which are chemically active result in the disfiguration of monuments. Intense solar radiation, moisture and changing climatic conditions lead to the decay of building materials used in monuments. Stone monuments, such as those made of limestone, are withering away. There is growth of algae, fungi, lichen, moss and higher plants on various monuments which make them esthetically unpleasant and also cause physical and chemical damage to the material. The oxidation of various metals used in the construction of the monument also causes monument's deformation and decreases its esthetic look. The conservation process involves the restoration and protection utilizing methods that effectively can keep the structure in the condition it was in originally for a prolonged period of time (Walston 1978). Conservation incorporates simple guidelines:

1. Minimalistic intervention.
2. Usage of appropriate materials and methods that are reversible.
3. Complete documentation of all undertaken work.
4. Reversibility.

The need for conservation of monuments was realized, and several methods were devised for their conservation and preservation. These methods are traditional or conventional and do not make use of advanced technologies. Since traditional building materials incorporated clay, earth, lime and mud, clay mortar is the oldest binding material that is used for the preservation of monuments. It confirms the longevity of monuments and is thus widely used for repair. The traditional methods like washing, water jet, cleaning, etc. proved to be not very efficient and often leave the monument damaged, with the damages being irreversible. It was thus essential to innovate methods to restore and conserve monuments which would be effective and which would not cause irreversible damage to the monument. This is where biotechnology steps in.

Biotechnology is a branch of science dealing with the technological application of living cells, organisms and systems to make or develop products. It harnesses biomolecular and cellular processes to innovate products and technologies that improve the standard of living and the health of the planet. Biotechnology is gaining popularity in almost all fields of life including healthcare, crop production and agriculture, industry and environment. Scientists have now started looking into biotechnology as an efficient way to restore and conserve monuments. Various innovative methods have been discovered such as biocleaning and the use of various bacteria and other microorganisms. The incorporation of biotechnology in the process has proven to be a changing step and can be a great tool in the systematic restoration and, therefore, protection of cultural heritage.

In this chapter, we will be looking at some of the causes of deterioration of monuments and will look upon a comparison between the traditional methods and the methods evolved using biotechnology used for the restoration and conservation of the monuments. We will first look at the traditional methods used followed by the application of biotechnology to restore the same deterioration method.

9.2 Causes of Deterioration

Causes of deterioration of monuments can largely be divided into two types, intrinsic causes which are strictly connected to the origin and nature of the building and extrinsic causes that are derived from external sources. The action of intrinsic causes is determined independently from the origin or from the different phases of the historical building's construction. The intrinsic causes can be further subdivided into two main subgroups, causes related to the position of the building and causes related to its structure, which is more common. The causes related to the structure are generally related to the natural and artificial materials used in the construction of the monument.

Extrinsic causes are those events that may have occurred subsequently or that remained to disturb the building. Extrinsic causes are further divided into two main groups, the forces of nature and those by human actions. There is a wide range of natural causes, and it is therefore convenient to divide extrinsic causes into three subgroups according to the speed of action.

1. Natural causes: prolonged action
2. Natural causes: occasional action
3. Causes provoked by human activity

The natural causes of occasional action include all the exceptional natural disaster phenomenons which are unpredictable, violent and difficult to avoid such as earthquakes, tsunamis and other natural disasters. The causes evoked by human activity include all the deliberate modifications in the original to the various structures, to the functions of the buildings and the changes in the surrounding environment and underground conditions. It also includes damages due to wars caused by human activities. Natural causes of prolonged action include the whole spectrum of chemical, electrochemical, physical, biological, botanical, microbiological, etc. actions that slowly sap the life of the building in all its structures and that can be classified under the heading of 'Ageing of the Construction'. The physical actions include all the damages caused to the structure by heat, water, wind and earth. It includes the damages caused by fluctuations of temperature, especially of temperature extremes due to fire and very low temperatures caused by freezing. The combined action of temperature and humidity favours chemical decay. Varieties of chemical reactions take place in the presence of humidity and are favoured by an increase in temperature. A rise in temperature and humidity also facilitates biological degradation by providing an excellent environment for the development of microorganisms and cryptogams (moulds, lichen, algae, etc.).

Botanical causes include ranges that are caused by the growth of plants, normal or parasitic vegetation. Plant growth in the immediate vicinity of historical buildings compromises conservation, especially when underground roots undermine the foundations and walls. Seeds often get deposited in the horizontal joints and projections. The fine roots penetrate the interior, grow slowly and act as wedges and separate and detach elements of the wall structure. Parasitic vegetation is more dangerous because it affects the facing and original surface of buildings like with poison ivy and other vines. It either aggravates the situation caused by normal plants or lives at the building.

Biological and microbial action includes all the damage caused by microorganisms (bacteria, etc.), often with accompanying biochemical transformations, as well as wood-eating insects (woodworms, white ants, termites) and rodents. The chemical reactions that result from the droppings of birds that nest or perch on historical buildings cause a lot of damage. Chemical and electrochemical causes include all the damages caused to the building by the interaction of various chemicals (from pollution) and water with the historical building. These include the damages caused by the accumulation of sulphur and its oxides (e.g. black crust of marble), nitrates and nitrites and other various chemicals (Fig. 9.1).

9.3 Biocleaning

Biocleaning is recognized as an excellent alternative to organic solvents and other chemical treatments and aggressive conservation methods such as mechanical treatment that have been used traditionally. Microorganisms can be used for conservation

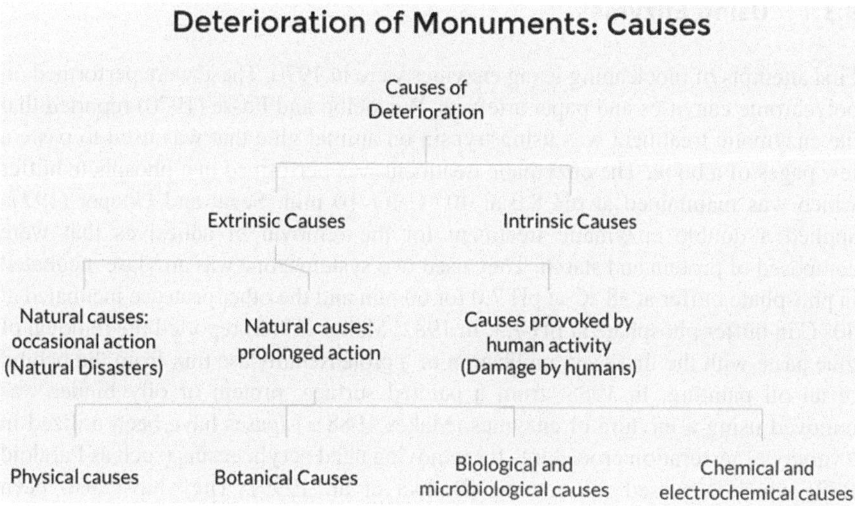

Fig. 9.1 Causes of deterioration of monuments

and restoration as bioagents on artworks and monuments. The basis of pioneering advancements of biological methods such as biocleaning is supported by studies that have shown that only a few known microorganisms can cause destruction (deterioration) naturally. Majority have been proved to be beneficial. Methods that use microorganisms have advantages over traditionally used chemical methods and enzymes, especially when the monuments under cleaning are encrusted and complex to clean, because of their specificity of the enzymes that they produce (Ranalli and Sorlini 2008). The research in the field of biocleaning technologies are leading to increase in the functionality in different varieties of environments, ranging from laboratories to various types of monuments. For the eradication of undesirable substances, a careful selection is made for the microorganisms (not pathogenic) that can be used that have the requisite characteristics. The undesirable substances include organic matter, nitrates, sulphates, etc. Microorganisms can be obtained from various places such as the international collections of microorganisms, or they can be isolated and collected from the soil or environmental matrices.

Biocleaning technologies have been applied to a number of monuments around the world including the Camposanto Monumentale cemetery in Pisa, Italy; the Duomo di Firenze Cathedral in Florence, Italy; the Duomo di Milano Cathedral in Milan, Italy; the Epidauro Theatre in Greece; Matera Cathedral in Matera, Italy; and the Santos Juanes Church in Valencia, Spain, artworks like Michelangelo's Pietà Rondanini and sculpture by Lina Arpesani and a building from the nineteenth century in Riga, Latvia.

9.3.1 Using Enzymes

First attempts of biocleaning using enzymes were in 1970. These were performed on polychrome canvases and paper artefacts. Wendelbo and Fosse (1970) reported that the enzymatic treatment was using trypsin on animal glue that was used to paste a few pages of a book. The enzymatic treatment was performed in a phosphate buffer which was maintained at pH 8.0 at 40 °C for 10 min. Segal and Cooper (1977) applied a double enzymatic treatment for the removal of adhesives that were composed of protein and starch. They used two systems: one was amylase incubated in phosphate buffer at 38 °C at pH 7.0 for 60 min and the other protease incubated at 40 °C in buffer phosphate at pH 7.5. In 1982, Makes (1982) reported the removal of glue paste with the first ever application of a protease/amylase mix from the behind of an oil painting. In 1988, from a painted surface, protein or oily binder was removed using a mixture of enzymes (Makes 1988). Lipases have been utilized in a variety of restoration procedures for removing aged acrylic resins (such as Paraloid B72) which are used for coating (Belluci et al. 1999). They have also been immobilized in an aqueous gel and used for the removal of oil-based overpaints (Vokic and Berovic 2005). Konkol and Macnamara (2009) demonstrated the decolouration using enzymes of spots generated on marble surfaces by the colouration effect of *Serratia* bacteria by laccase, a fungal enzyme (Konkol and Macnamara 2009).

Enzymes can be extracted from various biological sources. Amylase can be extracted from microbial sources (like *Bacillus* sp. or *Aspergillus* sp.) or animal tissues. Lipase can be derived from plant tissues (seeds of oats and wheat), from animal tissues (like pancreas) or from microbial species (such as *Bacillus*, *Aspergillus*, *Penicillium*). Pepsin, trypsin and protease can be extracted from both animal tissues (such as pancreas, stomach) and microbial cells (like *Bacillus*, *Aspergillus*).

9.3.2 Nitrate Salt Efflorescence

Wall paintings situated in indoor environments often have the formation of nitrate salt efflorescence on their surfaces which causes their major deterioration. Pressure is exerted, both on the wall and painted layers by the increased volume of crystals formed by accumulation of salts. Traction forces, produced due to the increasing pressure, tend to exceed the material strength, and eventually, micro-cracks are generated (Doménech-Carbó et al. 2006).

A Muslim mosque turned into a Baroque church in 1240, the Santos Juanes Church in Velencia was declared a National Monument of Historic and Artistic importance in 1942. It contains the largest wall paintings of Valencia. These cover the lunettes and the vault. The paintings, which were by artists Gulliό and Antonio Palomino, date back to between 1693 and 1702. They used the fresco technique in which pigments are applied using water and carbonation process is used for fixing to the lime and sand mortar.

9.4 Traditional Method for Restoration and Conservation

Physicochemical methods are usually used for the treatment of insoluble crusts on artworks. These methods are usually ineffective and can cause a lot of damage to the artwork. These methods require prolonged time of application and are aggressive, nonselective and invasive. They can also cause a change in colour, lead to movement of salt in the structure of the monument and cause the excessive removal of the original material. Toxic substances are used that pose a health risk to workers and lead to a release of toxic elements into the atmosphere which are not environment friendly (Cappitelli et al. 2007).

For the degradation of matter deposited that is organic, the use of *Pseudomonas stutzeri* has been reported (Antonioli et al. 2005; Ranalli and Sorlini 2003; Ranalli et al. 2005). Efflorescence of nitrate salts that developed on wall paintings placed in the central vault in lunettes in the church can be cleaned using *Pseudomonas stutzeri*. Used for removing nitrate salt crusts, *Pseudomonas stutzeri* is an example of nitrate-reducing bacteria. Molecular nitrogen is produced by the conversion of nitrates that were present in salts by using nitrate-reducing bacteria. Molecular nitrogen evaporates into the atmosphere since it is a gas at room temperature (Ranalli and Sorlini 2003).

Pseudomonas stutzeri can be applied to wall paintings by multiple methods assayed with Japanese paper behind them.

1. Sepiolite (Ranalli et al. 1997, 2000)
2. Cotton (Antonioli et al. 2005; Ranalli et al. 2005)
3. Carbogel (Cappitelli et al. 2006; Alfano et al. 2011)
4. Agar (Bosch-Roig et al. 2013)
5. Agarose (Bosch-Roig et al. 2013)

Stains can be caused by sepiolite and carbogel, confirming previously obtained results (Casanoba Beviá 2008). On surfaces that are horizontal, better results were observed by the application of cotton, agar and agarose (Bosch-Roig et al. 2010). Marks were observed to have been left by cotton on vertical application due to water leaks. Agar and agarose provide the best results, which due to the gravity effect of water produce a heterogeneously clean surface which contains a collection of bacteria towards the lower side. *P. stutzeri* was applied above which agar and Japanese paper were applied. This ensured that there was adequate humidity for the bacteria to work in. The Japanese paper and agar were removed after 90 min of treatment. Sterile water was used to delicately clean the treated wall surface. It was observed that the nitrate salt crusts had been completely eradicated after the painting surface had dried and no trace of *P. stutzeri* bacteria was left behind on the surface of the wall paintings (Fig. 9.2).

Fig. 9.2 (**a**) Isamu Noguchi's Slide Mantra. (**b**) Slide Mantra in Bayfront Park, Miami, Florida. The white Carrara marble sculpture stands 3 m high and weighs 29 tons. Red-brown stains near the base of the sculpture. (**c**) A pigmented bacterial culture (NRB) isolated from a red-brown stain grown on nutrient agar. (Konkol et al. 2009)

9.4.1 Black Crust of Marble

When calcium carbonate ($CaCO_3$) is exposed to the atmosphere unprotected, a chemical transformation occurs which converts it into calcium sulphate dehydrate, commonly referred to as gypsum (chemical formula, $CaSO_4{\cdot}2H_2O$), with the formation of sulphuric acid. The blackening of the sheltered areas is caused by the embedding of airborne pollutants like carbonaceous particles in the mineral matrix of gypsum during the process of crystallization. These blackened areas are referred to as black crusts of marble. Oxalates are formed in the crusts as a result of the microbial metabolism combined with the protective treatments (Sabbioni et al. 2003). Atmospheric pollution was believed to be the source of the anions of acetate and formate. The changes in the atmospheric pollution from time to time are depicted by the kind of pollutants in stones (Schiavon et al. 2004), and it was found that they are undergoing changes to show the pollutants in the atmosphere at that particular time. The black crust of marble not only is an esthetic problem but also causes damage to the material underneath.

A number of traditional procedures are in existence aimed towards eradication of the found 'black crust' on the surface of marble. One common method is the use of water sprays. Pressurised water jets or steam cleaning machines are used to remove gypsum from the areas where crust comes away from the marble and surface area without damaging it. The use of micro sandblasters is done in the areas where the crust does not come away as easily. These methods are very time-consuming and cause damage to the monument since they can cause the gypsum particles to enter the pores of the structure where they can later crystallize. They also make the surface relatively rough. Another method used is the use of a paste containing Komplexon Ill (EDTA). Komplexon binds with calcium and therefore also attacks the marble

($CaCO_3$). These methods are not completely effective. They produce uneven cleaning and do not result in the complete removal of crust from the marble surface. These damages can either be seen immediately, for example, the loss of surface or damages that might not be visible immediately or for years to come, in essence, the complex established microbiota be replaced by a single dominant species of micro-organism, post the removal in the cleaning process.

Skoulikidis and Beloyannis (1984) reported using aqueous solutions containing carbonate anions to produce calcite from gypsum found in the black crusts. The new layer formed, showing properties and behaviour similar to the underlying marble, was consolidated by calcite and the gypsum layer. By exploiting the specific metabolic processes of bacteria that reduce sulphur, the complete eradication of black crust (sulphate-based crusts) from the surfaces can be ensured. The studies on sulphate-reducing bacteria and their related metabolic processes began in 1989 (the earliest research being performed by Gauri et al. in 1989) dating up till Ranalli et al. in 1997 and Cappitelli et al. in 2006 and 2007. The bacteria used are *Desulfovibrio vulgaris*, *Desulfovibrio desulfuricans*, etc. The bacteria result in a progressive and uniform decrease in the black crust's thickness. Biotechnology for cleaning of black crust is preferred since bacterial cleaning can be controlled as it selectively removes specific compounds such as sulphates from damaged layers (Cappitelli et al. 2007), proving to be non-invasive and environment friendly.

Gypsum is dissociated into Ca_2C and SO_4^{2+} ions by the sulphate-reducing bacteria which are then further reduced. Carbon dioxide reacts with the Ca_2C ions to form the new calcite (Gauri and Bandyopadhyay 1999).

$$6CaSO_4 + 4H_2O + 6CO_2 = 6CaCO_3 + 4H_2S + 2S + 11O_2$$

Moncrieff and Hempel (1977) gave a description about the role of microorganisms in a cataplasm through the use of 'biological pack'. Atlas, Gauri and Chowdhury (Atlas et al. 1988) reported the first successful application of *Desulfovibrio desulfuricans*, an anaerobic sulphate-reducing bacterium. They obtained marble samples from a cornice of The Field Museum of Natural History, Chicago (Heselmeyer et al. 1991). *Desulfovibrio vulgaris* was applied to the crusts of gypsum in order to bring about the conversion into calcite. Gauri et al. in 1992 removed sulphate crusts using *D. desulfuricans* from the black crust on marble.

Similar works have been done on removal of black crust off of the Cathedral of Florence, Italy, including a marble column, made up of Carrara marble, which was obtained from under the dome of Brunelleschi in the Baccio D'Angelo Balcony and from an outer pilaster found in the Cathedral made out of stone typologies of three types, namely, white Carrara marble, red marlstone and green serpentine. The weathering of rocks to form soil is a degradation process that is essential for the continuation of life. It is caused by rain, wind, sunlight, moisture, temperature and snow. These affect the rock matrix's stability and cause dissolution of carbonates, oxidation and hydration reactions and solubilization of some elements which lead to chemical corrosion of minerals that form the stone (Koestler 2000). Additionally, anthropogenic factors (e.g. air pollution) lead to a higher concentration in organic and inorganic compounds in the atmosphere that eventually are deposited on stone surfaces of monuments and cause the exposed stone materials to decay. This weathering of rock caused the decay of culturally meaningful stone artefacts, buildings. etc. which represent an irreversible loss of cultural heritage.

Numerous preservation methods have been attempted towards the restoration of stone before irreversible damage is caused to it. Protection can be characterized as treatments that waterproof and additionally reinforce the stone surface so as to prevent the entry of water or other agents that cause weathering to the stone's core. The fortifying of and impregnation using solidifying product or cement of a friable rotting permeable stone is called consolidation. Traditionally, both the strategies have been performed on inorganic and organic materials, for example, Ba(OH)2 solution and acrylic/epoxy gums. Neither has demonstrated to be palatable. Organic treatments usually brought about the formation of incompatible, typically dangerous substances on the surface films which, by and large, result in the discharge of poisonous solvents. On the other hand, inorganic solidification would be favoured because of the presence of the common physicochemical fondness between the protective/consolidating materials and the stone minerals. For example, in the consolidation of carbonate stones, lime water treatment which contains $Ca(OH)_2$ solutions is used because $Ca(OH)_2$, in the presence of CO_2 in the atmosphere, easily carbonates and thereby results in calcite formation ($CaCO_3$) (De Muynck et al. 2010). However, this technique has been often found to lead to the accumulation of a friable, micrometre-thick, superficial, submicron-size aggregate of calcite crystals which result in the protection being insufficient.

The process of production of minerals by organisms is called biomineralization. The natural biogeochemical process in which inorganic materials are produced by microbes such as calcium carbonate precipitation, as a part of their daily metabolic activities, is called 'microbiologically induced calcium carbonate precipitation (MICP)'. The bacteria, supplied with the required liquid nutrients, are grown onto the affected stone and made to multiply and grow across the monument while they accumulate calcium. On the exhaustion of nutrient supply, they wither and dry off, leaving behind a coating rich in mineral.

MICP is regulated by four key factors as given by Hammes et al. (Hammes and Verstraete 2002):

1. The concentration of calcium
2. The concentration of dissolved inorganic carbon (DIC)
3. The pH
4. The nucleation sites' availability

MICP has been used in crack remediation. MICP has been examined for its adequacy explicitly in diminishing penetration properties and in prompting an improvement in strength of stone specimens by various researchers (Dick et al. 2006; Le Métayer-Levrel et al. 1999; Rodriguez-Navarro et al. 2003).

Sporosarcina pasteurii that performs MICP has been used to microbiologically enhance crack remediation. A new approach was taken towards encapsulating ureolytic bacteria cells in polyurethanes (PU) (Stocks-Fischer et al. 1999; Gollapudi et al. 1995). PU-immobilized bacterial cells report a high possibility to microbiologically enhance crack remediation. De Belie and De Muynck (2008) had used *Lysinibacillus sphaericus* for the calcite biomineralization for concrete crack repairs. Achal et al. (2013) reported the use of *Bacillus* sp. CT-5 in the bacterial calcite crack remediation.

A reduction of water absorptivity of the treated stone is observed by bacterially induced carbonates (Le Métayer-Levrel et al. 1999). Carbonates, phosphates and sulphates are precipitated by the application of *Myxococcus xanthus* (Rodriguez-Navarro et al. 2003) in a wide range of liquids and solids (Rodriguez-Navarro et al. 2007). Cementation using carbonates was observed to be effective up to a depth of a few hundred micrometres (>500 μm) with no discovered type of blocking or plugging of pores. Tiano et al. (1999) researched the consolidation effect on Pietra di Lecce limestone (bioclastic), through the utilization of *Micrococcus* sp. and *Bacillus subtilis*.

Zamarreño et al. (2009) researched the utilization of calcite crystals. In the specimens treated, the pore sizes of the substrate were found to have significantly reduced. De Muynck et al. (2013) reported that *Lysinibacillus sphaericus* in a range of temperatures (10, 20, 28, 37 °C) was found to be very effective for limestone consolidation. When applied to stones with two different types of pore sizes, i.e. microporous and macroporous, the success in macroporous stone was found to be more successful since the bacterial calcite application occurs to a larger depth and extent (De Muynck et al. 2011).

9.4.2 Red Stain on Marble

Red-, orange-, yellow- and brownish-purple-coloured stains appear to show on surfaces of marble monuments. These stains have been observed on several monuments of the world, including the Memorial Amphitheater; Arlington National Cemetery near Washington, D.C.; Fountain of Galatea at Villa Litta near Milan; the Labyrinth Fountain in Florence; and cathedral facades in Orvieto and Siena. These stains are caused by oxidation of lead or by a biological source (including red, heterotrophic bacteria of the genera *Micrococcus*, *Halococcus*, *Flavobacterium* and the red yeast *Rhodotorula minuta* and a few microorganisms that are photosynthetic). The Carrara marble, used in several Italian monuments, is stained by these bacteria.

Efforts have been made to remove the red stain on the monument installed in 1991 in Bayfront Park, Miami, Florida, Isamu Noguchi's Slide Mantra, which is made out of Carrara marble. The red stains on Slide Mantra were caused by a bacterium called *Serratia marcescens*. Traditional mechanical methods were used like cleaning with water sprays, air abrasion, etc. These methods can induce surface and/or structural damage to the monument (L. Lazzarini et al.). CB-4 biocidal detergent was also used for cleaning marble surface. Multiple CB-4 biocidal detergent treatments were applied to Slide Mantra's surface (Droycon Bioconcepts Inc.,

Regina, Saskatchewan). During extensive restoration work in early 2007, many a steam rinse treatments were employed. The treatment was successful to a certain extent in significantly removing all of the biological growth that is visible and in eliminating the surface stains. Uniquest CB-4 penetrates microbiological biofilms. The action of the detergent has been seen to result in three major impacts:

- The breaking apart and dispersion of biofilms
- The destruction of microbial cells, especially when concentrations applied exceed 0.5%
- Dispersing off of attached materials formed by the infestation and growth of microorganisms of the biofilm

The detergent removes surface soiling and leads to the exposure of the biofilm. It acts as a penetrating biocide and attacks the structures formed on and beneath stone surfaces. These structures collapse and can be easily removed by light scrubbing or rinsing. Cleaning of marble through enzymatic mechanisms can target the leftover reagents of discolouration and evidently restore the Slide Mantra and other such marble structures and artefacts resulting in very little to no damage to their structural esthetics. Plants such as those of olive and soybeanhave been used to extract white-rot fungi and lipoxidase (e.g. *Trametes versicolour*, and *Phanerochaete chrysospotium*) which have been used further to extract enzymes, for example, laccase, manganese and peroxidase, to be used in the paper and pulp industries and the food and textile industries to decolourize dyes and pigments. Laccase enzyme can be thus used in the restoration of cultural heritage sites (Fig. 9.3).

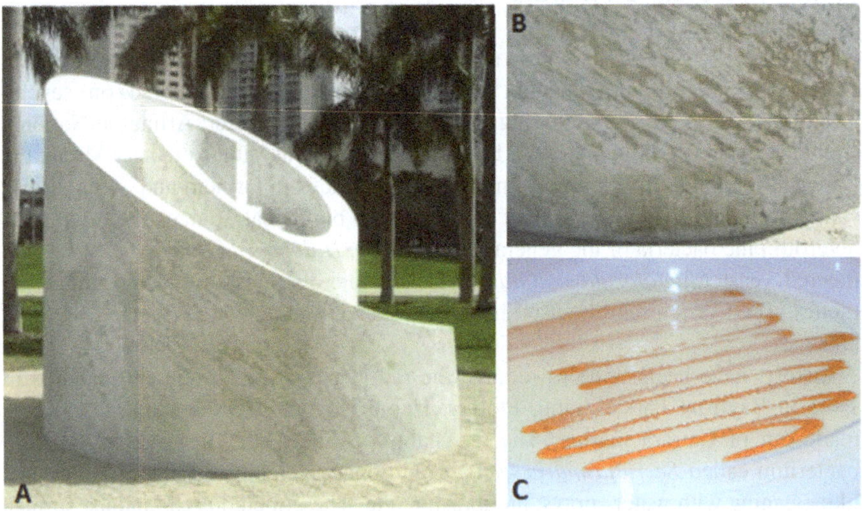

Fig. 9.3 Biocleaning with *Pseudomonas stutzeri* entrapped in an agar of salt efflorescence present on the wall painting lunette of the Santos Juanes Church of Valencia: (**a**) area before treatment, (**b**) area during treatment and (**c**) area after treatment. (Bosch-Roig et al. 2013)

A memorial made in honour of over 18,000 officers killed in the United States in the line of duty since 1792, the National Law Enforcement Officer's Memorial is made of Adair marble and contains two blue-grey walls, having inscribed the names of the martyred officers each surmounted by a group of lions protecting its younglings. It is set along with a variety of plants and trees and has accumulated stains and soils, consisting primarily of biological growth. The biological growth gets accumulated commonly in the pits or textured sections that tend to retain moisture, particularly on the horizontal surfaces, which are directly exposed to changing weather with little to no drainage or chance of drying. The soilage was more prominent in locations found under trees where there was a greater chance of combination of deposits from the trees, reduced rainwashing and shade, proving to be ideal conditions for the biological growth. Traditional cleaning tests proved unsuccessful in reducing the disfiguring. The paper poultice method for the transmission of CB-4, as it was done in the case of Slide Mantra, was used in an attempt to reduce the biological growth on the monument, but due to the highly porous Adair marble that is used in the monument, it was only partially effective, particularly on vertical surfaces where it tended to lose adhesion and pull away. Further, it did not inherently remove the killed biota afterwards and left nutrient-laden beds, thereby encouraging reinfestation and the return of bio-soiling.

The process was modified through the use of Prestór Gel (CBI Polymers, Honolulu, Hawaii, USA) as the transmission agent for CB-4 as a replacement for paper poultice. This enabled both the effective killing through prolonged contact with the biota of the biocide and the removal of the killed infestations from within the pits. Prestór Gel is based on a material developed to decontaminate nuclear sites. It is a polymeric gel and has the capacity to wick into microscopic pores. Unlike other commercially available gel cleaners (typically latex rubber-based that polymerizes into an elastic film), Prestór is water-soluble and water-based and offers much greater mechanical cleaning ability via its tenacious cling to and pull from the surface. It readily clings to vertical and overhead surfaces. Its relatively low viscosity necessitates careful site protection and application methods. With its drying, it shrinks by approximately 20% and results in the formation of a pliable, tough skin that encapsulates anything contained within it. After an approximate of 24 h, the gel can be peeled away and disposed of. This period is the necessary dwell time for the CB-4, meaning that it can be used to ensure that the CB-4 has sufficient contact time to act upon the bio-colonies effectively as well as remove it afterwards, therefore potentially reducing resoiling rates. One year after treatment, the stone of the Memorial remained unsoiled, thereby confirming its effectivity.

9.5 Damages by Biological and Microbial Action

Biodeterioration includes the damages caused by biological and microbial action. It is a global problem caused mainly by a variety of microorganisms observed on world heritage sites. Growth of microbes on stone material damages monuments which is often permanent and thus results in the irreversible loss of stone monuments by destroying them (Farooq et al. 2015). It accelerates the weathering by promoting severe surface pollution which further results in physical damage.

To prevent the monument biodeterioration, several traditional alternatives have been tested and tried to control microorganism growth. Some of these traditional methods include, but are not limited to, damp-proofing and waterproofing of surfaces; synthetic chemicals; leaf extracts in the aqueous state from plants possessing antimicrobial activity; potent ionic antimicrobial agents which are made from copper, zinc, silver, etc.; and inhibitors of activities of microorganisms (Ascaso et al. 2002). Since these traditional methods do not always lead to a positive impact and have at times led to harmful side effects, they are not preferred. They have proven to produce physical alterations and esthetic damage to colour and structural permeability or cause unfavourable chemical reactions such as solubilization of minerals or result in a change of pH of the monument (Peraza Zurita 2004). The penetration effect of chemicals is prevented by the microbial biofilms has been attributed to the failure of the chemical methods (Peraza Zurita 2004). These chemicals pose a threat to the environment and the public since they are toxic and hazardous. They can also adversely affect other organisms other than those targeted, for example, animals or humans, with some being carcinogenic and teratogenic, or induce chronic health problems when accumulated. Resistance to the chemicals can be acquired by some of the microbial species that remain growing on the walls.

9.5.1 Damages to Artwork

La Inmaculada, a piece that was infected and colonized by wood-eating insects, is a famous wood carving of an Immaculate, young Creole Virgin Mary (La Inmaculada), found in Venezuela.

Large-scale damage is found to be caused to historical artwork by insects, fungi and bacteria. Historical artworks are made up of a variety of material, one of the main materials used being wood and paper. Many insect species have been found to feed on paper and wood, the most common of these being termites belonging to the Isoptera taxonomic order Lepidoptera (carpenter worms) and Coleoptera (beetles).

In controlled museum environments, fungus control is made possible by reducing air humidity. Due to the high toxicity levels and possibility to harm artworks, biocides are not employed. Other physical traditional treatments such as heating, ultrasonic microwaves, high voltage and high frequency, electromagnetic waves, shocks of low current or freezing are not applied since the composition of a majority of wooden articles and paint substances gets affected by these treatments, thereby damaging the artwork. Oxygen depletion has also been proposed as a solution to the damage caused by insects. However, this method requires the use of costly equipment, and some of these insects (and their eggs) have been found to be able to survive for longer durations by switching to the anaerobic form of metabolism. Moreover, reinfestation by the same insect species is not prevented by oxygen depletion.

Alternatively, biological control involves the use of a special class of insects that are parasites on different species of insects called entomopathogenic fungi. They also include bacteria or viruses that possess the ability to kill bugs. The most common example of bacteria in this process is *Bacillus thuringiensis* (Bt). Toxins are produced that are capable of killing over a thousand different species of insects (Polanczyk and Alves 2003) including termites (Castilhos-Fortes et al. 2002). Bt cells contain the 'cry' protein, which is a protoxin which, in normal conditions, is highly insoluble. However, in the midgut of lepidopteran larvae, this gets solubilized mainly due to the reducing conditions and the presence of high pH (more than 9.5). It then gets activated after being cleaved and extracted by a protease from the gut, and the protoxin gets converted into the active form which is referred to as delta-endotoxin. This attaches to the epithelial cells of the midgut opening the cellular membrane pores, thereby disrupting the ion balance and, therefore, paralysing the digestive system. This results in starvation and septicaemia, thereby causing the death of the insects. Bt spores can be applied or sprayed onto the artwork. Bt spores have an added advantage over traditional methods. They tend to deteriorate easily, and no harmful side effects to the material are created. Furthermore, reinfestation by the same insects is prevented, thereby providing protection that is residual. Moreover, spores can resist harsh environments, and germination occurs only when in contact with the larvae of the insects. Moreover, spores can resist harsh environments and germination occurs only when in contact with the larvae of the insects ina very rich milieu which is found only in the hemolymph of the insect, thus making it impossible for them to grow and proliferate in paper or wood (Ramírez et al. 2006).

9.6 Biofilms in Restoration and Conservation

Biofilm is made up of any syntrophic consortium of possible microorganisms whose cells would stick to each other and to the surface of the material. The cells get embedded onto a slimy extracellular matrix made up of extracellular polymeric substance. They are capable of altering, colonizing and degrading a variety of materials including those which have been used for making cultural heritage monuments and artefacts like stone, paintings and frescoes (Sterflinger 2010). Biofilm has been reported on a large number of monuments across the globe including the South Korean Royal Tombs of the Joseon Dynasty (Yeoju, Yeongneung) and the Vat Phou and Associated Ancient Settlements of the Champasak Cultural Landscape (Vat Phou Temple). Biofilm on these monuments contained a wide variety of fungi and lichen. The number of fungal cultures isolated ranged from 11 to 21 species which included *Trichoderma* sp., *Penicillium* sp. and colonies that are unidentified (Jeong et al. 2018). *Bipolaris spicifera*, *Epicoccum nigrum*, *Aspergillus niger*, *Aspergillus ochraceus* and *Trichoderma viride* have also been observed.

In order to eradicate biofilms, several traditional methods have been employed such as use of biocides which combine quaternary ammonium compounds and benzalkonium chloride and ultraviolet radiation. The goal is to reduce concentrations of contaminants and preserve the quality and stability of the material. Biofilms that were formed on the stone monuments present in South Korea have been removed using chemical components with NA_2CO_3 and $(NH_4)_2CO_3$ as the main constituents. Globally, use of methods that involve the use of polishing and chemical treatments is not very recommended. Damages caused to the stone can be midterm to long term. The use of chemicals containing NA_2CO_3 and $(NH_4)_2CO_3$ has been associated with bringing about several problems. After the chemical application, distilled water is needed in a large quality, and safety of the workers is a great concern as well.

9.7 Essential Oils for Restoration and Conservation

In recent studies, the use of essential oils (EO) has been experimented upon for tackling the growth of fungi upon cultural heritage monuments. An essential oil is a natural oil obtained by distillation which is extracted from parts of various plants. It is a volatile chemical compound which is a hydrophobic liquid which is concentrated. They have a characteristic odour and have a variety of uses in various types of industries.

Essential oils extracted from aromatic plants belonging to the family Lamiaceae include *Lavandula angustifolia*, *Origanum vulgare* and *Rosmarinus officinalis*. Essential oils of various other kinds like cinnamaldehyde (95%, from cinnamon sp.), citral (95%, from lemongrass sp.), citronellol (95%, from rosae sp.), cuminaldehyde (98%, from cumin sp.), eucalyptol (99%, from eucalyptus sp.), eugenol (99%, from clove sp.), limonene (95%, from citric plant sp.), linalool (97%, from coriander sp.), menthol (99%, from mint sp.) and thymol (99% purity, main constituent of thyme EO) were also experimented upon, but the observed results for all were not as desirable. Only thymol, eugenol and cinnamaldehyde were able to restrict the fungal growth properly and for a long time.

These essential oils act as antifungal agents. Antifungal agents have the capacity to disrupt with any stage of the asexual life cycle of fungi. They not only act upon removing the previously occurred growth but also work upon inhibiting any further growth. The only observed problem with some of the essential oils is that they tend to evaporate since they are volatile and evaporate easily at room temperature. Essential oils have been used to remove biofilm layers from paintings, frescoes and walls of monuments. They have been used to eradicate fungal growth on Roman mural painting in Pompeii, Italy.

9.8 Eugenol Biocide for Restoration and Conservation

Eugenol is extracted from clove oil and is an allyl chain-substituted guaiacol. Natural biocide was made mainly of eugenol. The eugenol was identified and isolated from volatile clove extracts, Tween and Span series emulsifiers and water. Discolouration and decolourization of the biofilm was caused by the eugenol biocide. The biofilm gets degraded from monument's surface by the disintegration of the mycelium epithelial layer. Eugenol biocide is eco-friendly and non-toxic. An approximate 80% removal was seen on the surface to which the biocide was applied by calculating the measurement of ATP pre- and post-eugenol treatment. The use of eugenol biocide had several advantages over the traditional methods. The use of biocides did not result in a change in colour. It does not result in a change of weight either, which implies that the use of biocides does not cause any secondary weathering or damage to the stone material.

9.9 Silver Nanoparticle for Restoration and Conservation

Studies on silver have proved it to be a strong antimicrobial agent since many years. Silver is environmentally safe, has low toxicity to human cells and therefore is safe and has thermal and chemical stability (Bellissima et al. 2014). With the coming in of advanced technologies such as nanotechnology, it has been discovered that distinctive physicochemical properties are possessed by nanoparticles (NP) with sizes less than 100 nanometre. These particles could help solve several medical, agricultural and environmental problems in the world. Silver nanoparticles (AgNP) show high reactivity and have been proven to have an effective action in countering the growth and metabolism of a wide range of microorganisms such as bacteria, etc. (Choi et al. 2008).

The commercialization and public use of AgNP is due to the advancement of research on its use on biological systems (Lee et al. 2007). Experiments using AgNP on different strains of bacteria such as *E. coli* (Sondi and Salopek-Sondi 2004), termites (Kartal et al. 2009), fungi that are pathogenic to plant (Kim et al. 2006), *Cladosporium* growth on gypsum (Shirakawa et al. 2013), molds (Pietrzak et al. 2015a, b) and *Aspergillus niger* (Goffredo et al. 2017) have proven that AgNP when used as it is or in the combined compound state either with copper (Ruparelia et al. 2008; Coditã et al. 2010) or with titanium dioxide (Hamal et al. 2010; Li et al. 2011), has great biocidal effect on heterotrophic microorganisms. Bacterial, fungal and animal cells can be penetrated by AgNP (Navarro et al. 2008; Vass et al. 2015). A biochemical cascade gets activated after the AgNP interferes with the membrane proteins, and this activates cell division (McShan et al. 2014). The concentration of reactive oxygen species (ROS) increases after the entry of nanoparticles by endocytosis or diffusion through the plant cell wall and plant plasma membrane leading to mitochondrial dysfunction. As a result, ROS damages the surrounding nucleic acid and proteins by causing oxidative stress (Haase et al. 2012). The mobility inside the environment and biological system are a result of the amount of organic and inorganic compound interaction, size, and the toxic capacity of AgNP. Many environmental parameters, which are humidity, pH and temperature, also affect the toxicity of AgNP. In the maximum cases, degradation of chloroplast by AgNP treatment is dose and time dependent which causes the photosynthetic activity of the cells to decrease resulting in an inhibition of aerial algae growth. Restoration work on the pre-Hispanic city complex walls using AgNP has been done.

9.10 Major Challenges

Although methods of restoration and conservation utilizing biotechnology are environmentally friendly, safer and more effective, there are still several challenges that are obstructing its wide-scale use.

The risk associated with biotechnology is high. The methods need to be innovated, and this requires heavy funding. Several rounds of testing are conducted on various types of samples obtained from different sources. The raw materials used in these testings are expensive. The laboratories in which these testings are conducted need to be equipped with the proper machines and equipment which are also very expensive. Moreover, there is no guarantee that the outcome will be positive and most times the experiments fail. Furthermore, the advancements in science and technology are so rapid that there could occur major technological changes during the research and development phase that could have a negative impact on the product.

Most times the final product is also very expensive. This makes it unaffordable for large-scale use in many countries and regions of the world.

The use of biotechnology for restoration and conservation is very minimal and is restricted to a very few monuments in a few places in the world. Many of the methods that are being experimented upon get stuck for a long time within the laboratory itself to check the efficiency and validity of the technique being developed. The testing and commercialization of the conservation methods has also not been done in an efficient manner, therefore making it all the more difficult to spread the implementation. Moreover, many of the methods that have been discovered and experimented upon need further work to be done on them in order to make them completely functional and efficient (Table 9.1).

9.11 Conclusion and Future Prospects

The experiences with the use of biotechnology in the restoration and conservation of various monuments quoted in this chapter illustrate the utilization of biotechnology in the better conservation of monuments. Biotechnology is developing rapidly as field of science, and with the rapid advancements in science and technology, the growth and development in the techniques used in biotechnology are expected to advance a million fold. Biotechnological techniques are far more powerful than mechanical traditional and chemical methods. They are low cost and environmentally friendly and pose less risk to human health. By the integration of the expertise of various fields, novel and better techniques can be developed. The successes described in this chapter highlight the importance to fortify the communication among the worlds of science and culture so that biotechnology and other techniques based in the fields of chemistry and physics can be applied to the restoration works.

The methods in biotechnology that are being currently used, although effective, are still not being used at a large scale. Many already discovered methods have not been implemented in many countries of the world. Many methods are still being

Table 9.1 List of microorganisms that can be used for different types of deterioration

S.N.	Organisms	Phylum	Role/applications	References
1	*Bacillus amyloliquefaciens*	Firmicutes	Conservation of ornamental stones	Lee (2003)
2	*Bacillus cereus*	Firmicutes	Biocement (biocalcin) formation and limestone consolidation	Orial et al. (1993) and Casteneir et al. (2000)
3	*Bacillus lentus*	Firmicutes	Biocement (biocalcin) formation and limestone consolidation	Dick et al. (2006) and Sarda et al. (2009)
4	*Bacillus megaterium*	Firmicutes	Calcite precipitation	Cacchio et al. (2003)
5	*Bacillus pasteurii*	Firmicutes	Biocement formation and consolidation of sand column and repair of concrete cracks	Sarda et al. (2009)
6	*Bacillus pumilus*	Firmicutes	Calcite precipitation	Baskar et al. (2006)
7	*Bacillus sphaericus*	Firmicutes	Concrete consolidation Biocement (biocalcin) formation	Muynck et al. (2008) and Dick et al. (2006)
8	*Bacillus thuringiensis*	Firmicutes	Calcite precipitation	Baskar et al. (2006)
9	*Bacillus subtilis*	Firmicutes	Reduction in water absorption of 'Pietra di Lecce', a calcareous stone, thereby rendering consolidation	Tiano et al. (1999)
10	*Desulfovibrio desulfuricans*	Proteobacteria	Removal of black crust from marble surface	Gauri et al. (1992)
11	*Desulfovibrio vulgaris*	Proteobacteria	Removal of black crust from marble surface	Cappitelli et al. (2007)
12	*Halobacillus trueperi*	Proteobacteria	Biomineralization	Rivadeneyra et al. (2004)
13	*Lysinibacillus sphaericus*	Firmicutes	Calcite precipitation	De Belie and De Muynck (2008)
14	*Morchella* sp.	Ascomycetes	Formation of carbonate concretions	Masaphy et al. (2009)
15	*Myxococcus xanthus*	Proteobacteria	Consolidation of ornamental stones Bioconservation of cultural heritage structures	Rodriguez-Navarro et al. (2003)
16	*Pseudomonas cepacia*	Proteobacteria	Biocleaning, restoration and removal of animal glue from fresco	Ranalli et al. (2005)
17	*Pseudomonas fluorescens*	Proteobacteria	Biocleaning and restoration of fresco calcite precipitation	Anderson et al. (1992) and Ranalli et al. (2005)

(continued)

Table 9.1 (continued)

S.N.	Organisms	Phylum	Role/applications	References
18	*Pseudomonas halophila*	Proteobacteria	Calcite precipitation	Rivadenerya et al. (2006)
19	*Pseudomonas flavescens*	Proteobacteria	Biocleaning, restoration and removal of animal glue from fresco	Ranalli et al. (2005)
20	*Pseudomonas stutzeri*	Proteobacteria	Biocleaning, restoration and removal of animal glue from fresco	Ranalli et al. (2005)
21	*Sporosarcina pasteurii*	Firmicutes	Calcite precipitation	Gollapudi et al. (1995) and Stocks-Fischer et al. (1999)
22	*Rhodococcus erythropolis*	Actinobacteria	Consolidation of 'Pietra di Lecce', a calcareous stone	Sprocati et al. (2008)
23	*Thiobacillus* sp.	Proteobacteria	Removal of fouled layer of lichen from weathered concrete specimens	De Graef et al. (2005)
24	*Pseudomonas* and *Bacillus* sp.	Proteobacteria	Removal of phenanthrene deposit from weathered stones	Saiz-Jimenez (1997)
25	*Pseudomonas* sp.	Proteobacteria	Removal of nitrates from the weathered stones	Ranalli et al. (1997)
26	*Pseudomonas* sp.	Proteobacteria	Limestone consolidation	Zamarreño et al. (2009)
27	*Acinetobacter*	Proteobacteria	Limestone consolidation	Zamarreño et al. (2009)
28	*Micrococcus*	Actinobacteria	Reduction in water absorption of 'Pietra di Lecce', a calcareous stone, thereby rendering consolidation	Tiano et al. (1999)

researched upon in labs. One such method is the use of silver nanoparticles in the restoration against damages by biological and microbial growth. Silver nanoparticles hold great potential in restoration against damages by biological and microbial growth, but their use is next to nothing. The use of biotechnology for the restoration and conservation of monuments is very little in India. Major monuments of India like the Taj Mahal in Agra are being deteriorated by insects and pollution. These insects can be tackled by various innovations of biotechnology like Bt. Biofilms have also been observed on the Taj Mahal. These can be eradicated using essential oils instead of harmful chemicals. Various other monuments in India are also facing various types of deterioration. The Kutub Minar in Delhi has now tilted due to a weak foundation and rainwater seepage. The Lotus Temple in Delhi is also withering under the growing pollution. The Kutub Minar and the Lotus Temple could perhaps be restored by the use of the process of biocementation. Many other monuments are getting cracks because of pollution and ageing. These can also be repaired by the process of biomineralization.

Francesca Cappitelli, a microbiologist working on art conservation at the University of Milan, Italy, is of the opinion that 'In the near future, the expanding frontiers of biotechnology in conservation will include, as a few examples, the control of quorum sensing to avoid the formation of biofilms on cultural heritage objects and the use of single-cell biosensors for the museum environment monitoring and control'. It is essential that the world use biotechnology as the way for restoration and conservation of monuments. This will not only help reduce the harmful effects of chemicals but also open new horizons in art preservation and provide novel art-oriented careers for those in the field of biology.

The given chapter looks upon the restoration and conservation of monuments using biotechnology. There are a number of methods and techniques that have been devised for this purpose. Various strains of microorganisms are being used to tackle and eradicate other microorganisms. Biocleaning as a process has been discussed. The use of enzymes extracted from biotic sources has been discussed. Biotechnology is a much safer and environmentally friendly alternative and is also better for human health. It has proven to be a better restoration and conservation method as compared to chemical and mechanical methods and is more permanent as well. Advances in science and technology pave the way for the advancement of biotechnology and, therefore, better alternative methods for restoration and conservation.

References

Achal V, Mukerjee A, Sudhakara Reddy M (2013) Biogenic treatment improves the durability and remediates the cracks of concrete structures. Elsevier, Amsterdam

Alfano G, Lustrato G, Belli C, Zanardini E, Cappitelli F, Mello E, Sorlini C, Ranalli G (2011) The bioremoval of nitrate and sulfate alterations on artistic stonework: the case-study of Matera cathedral after six years from the treatment. Int Biodeterior Biodegrad 65(7):1004–1011

Anderson S, Appanna VD, Huang J, Viswanatha T (1992) A novel role for calcite in calcium homeostasis. FEBS Lett 308:94–96

Antonioli P, Zappararoli G, Abbruscato P, Sorlini C, Ranalli G, Righetti PG (2005) Art-loving bugs: the resurrection of Spinello Aretino from Pisa's cemetery. Proteomics 5:2453e2459

Ascaso C, Wierzchos J, Souzaegipsy V, Delosrios A, Rodrigues J (2002) In situ evaluation of the biodeteriorating action of microorganisms and the effects of biocides on carbonate rock of the Jeronimos Monastery (Lisbon). Int Biodeterior Biodegrad 49(1):1–12

Atlas RM, Chowdhury AN, Gauri LK (1988) Microbial calcification of gypsum–rock and sulphated marble. Stud Conserv 33:149–153

Baskar S, Baskar R, Mauclaire L, Mckenzie JA (2006) Microbially induced calcite precipitation in culture experiments: possible origin for Sahastradhara Caves, Dehradun, India. Curr Sci India 90:58–64

Bellissima F, Bonini M, Giorgi R, Baglioni P, Barresi G, Mastromei G, Perito B (2014) Antibacterial activity of silver nanoparticles grafted on stone surface. Environ Sci Pollut Res 21:13278–13286

Bellucci R, Cremonesi P, Pignagnoli G (1999) A note on the use of enzymes in conservation. A preliminary report on the removal of aged acrylic resin coatings with Lipase. Stud Conserv 44:278–281

Bosch-Roig P, Regidor-Ros JL, Soriano-Sancho P, Doménech-Carbó MT, Montes- Estellés RM (2010) Ensayos de Biolimpieza con Bacterias en Pinturas Murales. Arché 4e5:115e122
Bosch-Roig P, Regidor-Ros J, Soriano P, Montes-Estelles R (2013) Biocleaning of animal glue on wall paintings by *Pseudomonas stutzeri*. Chimica Oggi/Chem Today Elsevier 31:50–53
Cacchio P, Ercole C, Cappuccio G, Lepidi A (2003) Calcium carbonate precipitation by bacterial strains isolated from limestone cave and from a loamy soil. Geomicrobiol J 20:85–98
Cappitelli F, Zanardini E, Ranalli G, Mello E, Daffonchio D, Sorlini C (2006) Improved methodology for bioremoval of black crusts on historical stone artworks by use of sulfate-reducing bacteria. Appl Environ Microbiol 72:3733–3737
Cappitelli F, Toniolo L, Sansonetti A, Gulotta D, Ranalli G, Zanardini E et al (2007) Advantages of using microbial technology over traditional chimica technology in removal of black crust from stone surfaces of historical monuments. Appl Environ Microbiol 73:5671–5675
Casanoba Beviá I (2008) Limpieza de obra mural mediante el uso de bacterias. Master thesis, Polytechnic University of Valencia, Spain
Castanier S, Le Métayer-levrel G, Orial G, Loubìere JF, Perthuisot JP (2000) Bacterial carbonatogenesis and applications to preservation and restoration of historic property. In: Of microbes and art: the role of microbial communities in the degradation and protection of cultural heritage. Plenum, New York, pp 201–216
Castilhos-Fortes R et al (2002) Susceptibility of *Nasutitermes ehrhardti* (Isoptera: Termitidae) to *Bacillus thuringiensis* subspecies. Braz J Microbiol 33:219–222
Choi O, Deng KK, Kim NJ, Ross L Jr, Surampallie RY, Hu Z (2008) Inhibitory effects of silver nanoparticles, silver ions, and silver chloride colloids on microbial growth. Water Res 42:3066–3074
Codiţă I, Caplan DM, Drăgulescu EC, Lixandru BE, Coldea IL, Dragomirescu CC et al (2010) Antimicrobial activity of copper and silver nanofilms on nosocomial bacterial species. Roum Arch Microbiol Immunol 69:204–212
De Belie N, De Muynck W (2008) Crack repair in concrete using biodeposition. In: Concrete repair, rehabilitation and retrofitting II. CRC Press, The Netherlands, pp 309–310
De Graef B, De Windt W, Dick J, Verstraete W, De Belie N (2005) Cleaning of concrete fouled by lichens with the aid of Thiobacilli. Mater Struct 38:875–882
De Muynck W, Debrouwer D, De Belie N (2008) Bacterial carbonate precipitation improves the durability of cementitious materials. Cem Concr Res 38:1005–1014
De Muynck W, De Belie N, Verstraete W (2010) Microbial carbonate precipitation in construction materials: a review. Ecol Eng 36:118–136
De Muynck W, Leuridan S, Van Loo D, Verbeken K, Cnudde V, De Belie N, Verstraete W (2011) Influence of pore structure on the effectiveness of a biogenic carbonate surface treatment for limestone conservation. Appl Environ Microbiol 77(19):6808–6820
De Muynck W, Verbeken K, De Belie N, Verstraete W (2013) Influence of temperature on the effectiveness of a biogenic carbonate surface treatment for limestone conservation. Appl Microbiol Biotechnol 97(3):1335–1347
Dick J, De Windt W, De Graef B, Saveyn H, Van Der Meeren P, De Belie N, Verstraete W (2006) Bio-deposition of a calcium carbonate layer on degraded limestone by Bacillus species. Biodegradation 17:357–367
Doménech-Carbó MT, Yusá-Marco DJ (2006) Aspectos físicoquímicos de la pintura mural y su limpieza. Polytechnic University of Valencia, Valencia
Farooq M (2015) Mycobial deterioration of stone monuments of dharmarajika, taxila. J Microbiol Exp 2(1):29–33
Gauri KL, Bandyopadhyay JK (1999) Carbonate stone, chemical behaviour, durability, and conservation. Wiley, Chichester/New York
Gauri KL, Chowdhury Ahad N, Kulshreshtha Niraj P, Punur Adinarayana R (1989) The sulfation of marble and the treatment of gypsum crusts. Stud Conserv 34:201–206

Gauri KL et al (1992) Removal of sulfated-crust from marble using sulfate reducing bacteria. In: Webster RGM (ed) Stonecleaning and the nature, soiling and decay mechanisms of stone. Donhead, London, pp 160–165

Goffredo GB, Citterio B, Biavasco F, Stazi F, Barcelli S, Munafò P (2017) Nanotechnology on wood: the effect of photocatalytic nanocoatings against *Aspergillus niger*. J Cult Herit 14:1–12. in press

Gollapudi UK, Knutson CL, Bang SS, Islam MR (1995) A new method for controlling leaching through permeable channels. Chemosphere 30(4):695–705

Haase A, Rott S, Mantion A, Graf P, Plendl J, Thünemann AF et al (2012) Effects of silver nanoparticles on primary mixed neural cell cultures: uptake, oxidative stress and acute calcium responses. Toxicol Sci 126:457–468

Hamal DB, Haggstrom JA, Marchin GL, Ikenberry MA, Hohn K, Klabunde KJ (2010) A multi-functional biocide/sporocide and photocatalyst based on titanium dioxide (TiO2) codoped with silver, carbon, and sulphur. Langmuir 26:2805–2810

Hammes F, Verstraete W (2002) Key roles of pH and calcium metabolism in microbial carbonate precipitation. Rev Environ Sci Biotechnol 1:3–7

Heselmeyer K, Fischer U, Krumbein WE, Warsheid T (1991) Application of *Desulfovibrio vulgaris* for the bioconversion of rock gypsum crusts into calcite. BIOforum 1:89

Jeong SH, Lee HJ, Kim DW, Chung YJ (2018) New biocide for eco-friendly biofilm removal on outdoor stone monuments. Int Biodeterior Biodegrad 131:19–28

Kartal SN, Green F, Clausen CA (2009) Do the unique properties of nanometals affect leachability or efficacy against fungi and termites. Int Biodeterior Biodegrad 63:490–495

Kim SW, Jung JH, Lamsal K, Kim YS, Min JS, Lee YS (2006) Antifungal effects of silver nanoparticles (AgNPs) against various plant pathogenic fungi. Mycobiology 40:53–58

Koestler RJ (2000) When bad things happen to good art. Int Biodeterior Biodegrad 46:259–269

Konkol N, Macnamara C (2009) Enzymatic decoloration of bacterial pigmented from culturally significant marble. J Cult Herit 10:362–366

Konkol N, McNamara C, Sembrat J, Rabinowitz M, Mitchell R (2009) Enzymatic decolorization of bacterial pigments from culturally significant marble. J Cult Herit 10:362–366

Le Métayer-Levrel G, Castanier S, Orial G, Loubière J-F, Perthuisot J-P (1999) Applications of bacterial carbonatogenesis to the protection and regeneration of limestones in buildings and historic patrimony. Sediment Geol 126(1-4):25–34

Lee YN (2003) Calcite production by Bacillus amyloliquefaciens CMB01. J Microbiol 41:345–348

Lee HY, Park HK, Lee YM, Park SB (2007) A practical procedure for producing silver nanocoated fabric and its antibacterial evaluation for biomedical applications. Chem Commun 28:2959–2961

Li M, Noriega-Trevino ME, Nino-Martinez N, Marambio-Jones C, Wang J, Damoiseaux R et al (2011) Synergistic bactericidal activity of Ag-TiO2 nanoparticles in both light and dark conditions. Environ Sci Technol 45:8989–8995

Makes F (1982) Enzymatic consolidation of a painting: seventeenth century landscape from Skokloster Palace. In: Contribution of the Washington congress, IIC, London, 25–30 October

Makes F (1988) Enzymatic consolidation of the portrait of Rudolf II with a multi-enzyme preparation isolated from Antartic krill, Goteborg studies on conservation 1. Acta Universitatis Gothoburgensis, Göteborg

Masaphy S, Zabari L, Pastrana J, Dultz S (2009) Role of fungal mycelium in the formation of carbonate concretions in growing media—an investigation by SEM and synchrotron-based X-ray tomographic microscopy. Geomicrobiol J 26:442–450

McShan D, Ray PC, Yu H (2014) Molecular toxicity mechanism of nanosilver. J Food Drug Anal 22:116–127

Moncrieff A, Hempel K (1977) Conservation of sculptural stonework: virgin & child on S. Maria dei Miracoli and the Loggetta of the campanile, Venice. Stud Conserv 22:1–11

Navarro E, Baun A, Behra R, Hartmann NB, Filser J, Miao AJ et al (2008) Environmental behaviour and ecotoxicity of engineered nanoparticles to algae, plants, and fungi. Ecotoxicology 17:372–386

Orial G, Castanier S, Le Metayer G, Loubière JF (1993) The biomineralization: a new process to protect calcareous stone applied to historic monuments. In: Ktoishi H, Arai T, Yamano K (eds) Proceedings of the 2nd international conference of biodeterioration of cultural property. International Communications Specialists, Tokyo, pp 98–116

Peraza Zurita Y (2004) Biodeterioro por microalgas en fuentes de mármol. Universidad de Granadas, Ph.D. thesis

Pietrzak K, Gutarowska B, Machnowski W, Mikołajczyk U (2015a) Antimicrobial properties of silver nanoparticles misting on cotton fabrics. Text Res J 86:812–822

Pietrzak K, Twarużek M, Czyżowska A, Kosicki R, Gutarowska B (2015b) Influence of silver nanoparticles on metabolism and toxicity of moulds. Acta Biochim Pol 62:851–857. pmid:26637374

Polanczyk R, Alves S (2003) Bacillus thuringiensis: uma breve revisão. Agrociencia 7:1–10

Ramírez JL, Santana MA, Galindo-Castro I, Gonzalez A (2006) The role of biotechnology in art preservation. Trends Biotechnol 23:584–588

Ranalli G, Sorlini C (2003) Application of microorganisms for the deteriorates surfaces recovery. Coalition 6:2e4

Ranalli G, Sorlini C (2008) Bioremediation. In: Caneva G, Nugari MP, Salvadori O (eds) Plant biology for cultural heritage, Biodeterioration and conservation. The Getty Conservation Institute, Los Angeles

Ranalli G, Chiavarini M, Guidetti V, Marsala F, Matteini M, Zanardini E et al (1997) The use of microorganisms for the removal of sulphates on artistic stoneworks. Int Biodeterior Biodegrad 40:255–261

Ranalli G, Matteini M, Tosini I, Zanardini E, Sorlini C (2000) Bioremediation on cultural heritage: removal of sulphates, nitrates and organic substances. In: Ciferri O, Tiano P, Mastromei G (eds) Of microbes and art: the role of microbial communities in the degradation and protection of cultural heritage. Kluwer Academic-Plenum, New York, p 231e245

Ranalli G, Alfano G, Belli C, Lustrato G, Bonadduce I, Colombini MP, Zanardini E, Abbruscato P, Cappitelli F, Sorlini C (2005) Biotechnology applied to cultural heritage: biorestoration of frescoes using viable bacterial cells and enzymes. J Appl Microbiol 98:73e83

Rivadeneyra MA, Párraga J, Delgado R, Ramos-cormerzana A, Delgado G (2004) Biomineralization of carbonates by *Halobacillus trueperi* in solid and liquid media with different salinities. FEMS Microbiol Ecol 48:39–46

Rivadeneyra MA, Delgado R, Párraga J, Ramos-cormenza A (2006) Precipitation of minerals by 22 species of moderately halophilic bacteria in artificial marine salts media: influence of salt concentration. Folia Microbiol 51:445–453

Rodriguez-navarro C, Rodriguez-gallego M, Chekroun KB, Gonzalez-muñoz MT (2003) Conservation of ornamental stone by *Myxococcus xanthus*-induced carbonate biomineralization. Appl Environ Microbiol 69:2182–2193

Rodriguez-Navarro C, Jimenez-Lopez C, Rodriguez-Navarro A, Gonzalez-Muñoz MT, Rodriguez-Gallego M (2007) Bacterially mediated mineralization of vaterite. Geochim Cosmochim Acta 71(5):1197–1213

Ruparelia JP, Chatterjee AK, Duttagupta SP, Mukherji S (2008) Strain specificity in antimicrobial activity of silver and copper nanoparticles. Acta Biomater 4:707–716. pmid:18248860

Sabbioni C et al (2003) Organic anions in damage layers on monuments and buildings. Atmos Environ 37:1261–1269

Saiz-jimenez C (1997) Biodeterioration vs biodegradation: the role of microorganisms in the removal of pollutants deposited on historic buildings. Int Biodeterior Biodegrad 40:225–232

Sarda D, Choonia HS, Sarode DD, Lele SS (2009) Biocalcification by Bacillus pasteurii urease: a novel application. J Ind Microbiol Biotechnol 36:1111–1115

Schiavon N et al (2004) Soiling of limestone in an urban environment characterized by heavy vehicular exhaust emissions. Environ Geol 46:448–455

Segal J, Cooper D (1977) The use of enzymes to release adhesives. Pap Conserv 2:47–50

Shirakawa MA, Gaylarde CC, Sahão HD, Lima JRB (2013) Inhibition of Cladosporium growth on gypsum panels treated with nanosilver particles. Int Biodeterior Biodegrad 85:57–61

Skoulikidis TN, Beloyannis N (1984) Inversion of marble sulfation: reconversion of gypsum films into calcite on the surfaces of monuments and statues. Stud Conserv 29:197–204

Sondi I, Salopek-Sondi B (2004) Silver nanoparticles as antimicrobial agent: a case study on *E. coli* as a model for gram-negative bacteria. J Colloid Interface Sci 275:177–182

Sprocati AR, Alisi C, Tasso F (2008) A microbial survey of the Etruscan Mercareccia Tomb (Italy): contribution of microorganisms to deterioration and restoration. In: 9th international conference on NDT of Art, Jerusalem, Israel

Sterflinger K (2010) Fungi: their role in deterioration of cultural heritage. Fungal Biol Rev 24 (1–2):47–55

Stocks-Fischer S, Galinat JK, Bang SS (1999) Microbiological precipitation of CaCO3. Soil Biol Biochem 31(11):1563–1571

Tiano P, Biagiotti L, Mastromei G (1999) Bacterial bio-mediated calcite precipitation for monumental stones conservation: methods of evaluation. J Microbiol Methods Elsevier 36:139–145

Vass IZ, Deák Z, Paul K, Kovács S, Vass I (2015) Interaction of nanoparticles with biological systems. Acta Biol Szeged 59:225–245

Vokic D, Berovic M (2005) Use of lipase to remove oil-base overpaints. ICOM- 13th Triennal meeting, 12–16 September

Walston S (1978) The preservation and conservation of aboriginal and Pacific cultural material in Australian museums. ICCM Bull 4(4):9–21

Wendelbo O, Fosse B (1970) Protein surgery: a restoring procedure applied on paper. Restaurator 1:245–248

Zamarreño DV, May E, Robert I (2009) Influence of environmental temperature on biocalcification by non-sporing freshwater bacteria. Geomicrobiol J 26:1–12

Biocement: A Novel Approach in the Restoration of Construction Materials

10

Hesham El Enshasy, Daniel Joe Dailin, Roslinda Abd Malek, Nurul Zahidah Nordin, Ho Chin Keat, Jennifer Eyahmalay, Santosh Ramchuran, Jimmy Ngow Chee Ghong, Veshara Malapermal Ramdas, and Rajesh Lalloo

Abstract

Concrete is the most commonly used construction material worldwide for the development of durable structures. Structural integrity and design of buildings have become increasingly important in construction engineering as well as assessment of mixed formulation including cement and aggregate (i.e. sand, slag and stone). Microcrack formation on concrete may result in increased

H. El Enshasy (✉)
Institute of Bioproduct Development (IBD), Universiti Teknologi Malaysia (UTM), Johor Bahru, Malaysia

School of Chemical Engineering and Energy, Faculty of Engineering, Universiti Teknologi Malaysia (UTM), Johor Bahru, Malaysia

Genetic Engineering and Biotechnology Research Institute, City of Scientific Research and Technology Applications (CSAT), New Burg Al Arab Alexandria, Egypt
e-mail: henshasy@ibd.utm.my

D. J. Dailin · R. A. Malek · N. Z. Nordin · J. Eyahmalay
Institute of Bioproduct Development (IBD), Universiti Teknologi Malaysia (UTM), Johor Bahru, Malaysia

School of Chemical Engineering and Energy, Faculty of Engineering, Universiti Teknologi Malaysia (UTM), Johor Bahru, Malaysia

H. C. Keat
Institute of Bioproduct Development (IBD), Universiti Teknologi Malaysia (UTM), Johor Bahru, Malaysia

S. Ramchuran · R. Lalloo
Council for Scientific and Industrial Research (CSIR), Bioprocess Development Group, Pretoria, South Africa

J. N. Chee Ghong
Huashi Malaysia Sdn. Bhd, Kuala Lumpur, Malaysia

V. M. Ramdas
School of Life Sciences, University of KwaZulu Natal, Durban, South Africa

A. N. Yadav et al. (eds.), *Microbial Biotechnology Approaches to Monuments of Cultural Heritage*, https://doi.org/10.1007/978-981-15-3401-0_10

degradation and porous concrete. Therefore, there is a need to preserve and maintain concrete structures due to its high associated cost of restoration. In addition, reducing the negative environmental impact due to high CO_2 emissions during cement production need to be considered as well. One key solution includes bio-based self-healing techniques. Research has focused on biomineralisation, a method of sealing microcracks using bacterial calcium carbonate deposits, via a common process of biocementation or microbiologically induced calcium carbonate precipitation (MICP). As such, these deposits possess promising micro-bonding and pore-filling macro-effects for potential application in the construction industry. In view of these novel state-of-the-art techniques, this chapter provides an overview of potential microbes, mode of action of the self-healing process, primary limitations for future techniques and potential applications in the construction industry.

Keywords

Biomineralisation · Self-healing · Biocement · Calcium carbonate · Urease · Building materials

10.1 Introduction

Cement has an annual output of about 10 km^3/year and is also one of the most commonly used materials in the construction industry (Flatt et al. 2012; Pacheco-Torgal and Labrincha 2013; Sikder and Saha 2019). Cement is the only component produced, while other local materials are naturally accessible. One ton of cement manufacturing generates about one ton of CO_2 and constitutes 50% of worldwide CO_2 emissions into the atmosphere from the building sectors (Reddy et al. 2015). Concrete is a very suitable material to withstand the maximum compressive load or stress. However, if the load applied to the concrete exceeds its resistant load limit, the concrete's resistance decreases, thus forming cracks, unfortunately, with a very costly associated crack repair process. Concrete structures are highly dependent on the following properties, durability (resistance to stress) or resistance to permeability (i.e. water penetration), which may eventually come into contact with the concrete structure reinforcement resulting in corrosion and ultimately compromising the structural strength of the concrete (Pacheco-Torgal and Labrincha 2013; Luhar and Gourav 2015).

Therefore, there have been a number of approaches developed to increase the sustainability of concrete. The easiest method is to substitute (partially) cement with greener options, such as blast-furnace slag, which are iron and coal by-products (Knoben 2011). Another option is to increase concrete life, by reducing the need to replace materials. This is done now by regular inspection and structural restoration, but it is time-consuming and expensive. In addition, concrete degradation usually begins with microcracks that are barely visible to the naked eye. These cracks do not affect the strength of the framework themselves, but grow slowly. This contributes to

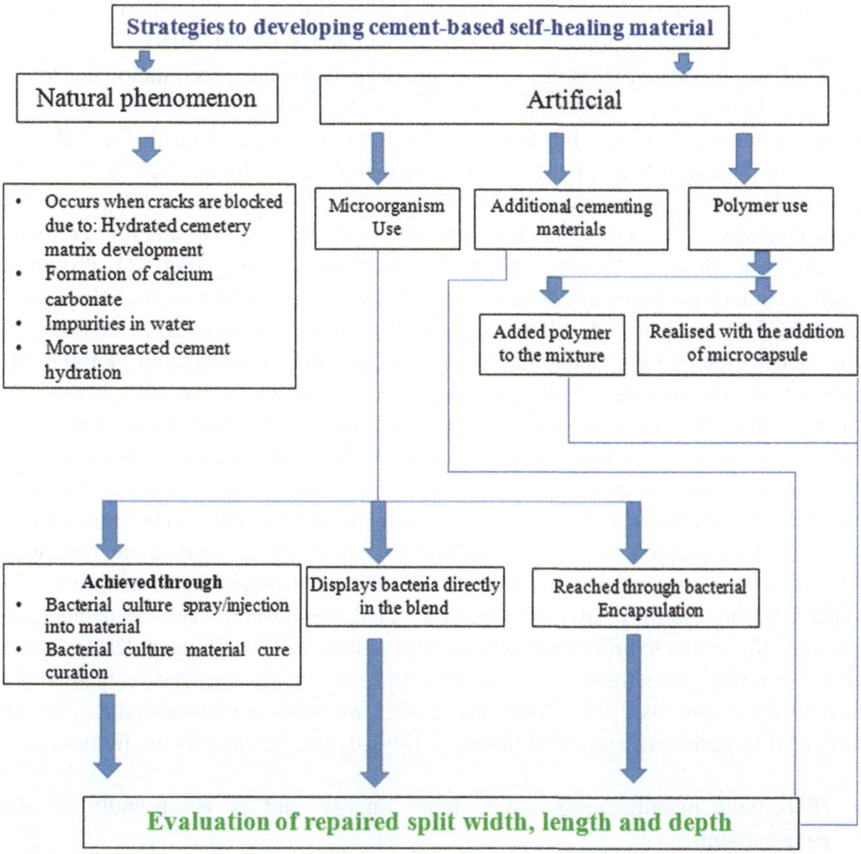

Fig. 10.1 A flow chart of the self-healing strategy in cement (Muhammad et al. 2016; Noshi and Schubert 2018)

the structure's corrosion and eventually failure. Compared to a normal cement, biocemetery has many benefits, for example, slightly distinct manufacturing processes in sandstone manufacturing, far shorter biocemetery time, appropriate for in situ processes, manufacturing of raw biological material at lower temperatures and more effective than normal cement, which in conventional manufacturing uses temperature up to 1500 °C.

Self-healing is therefore essential to strengthen the longevity of construction and cultural heritage systems (Schlangen et al. 2010). Several laboratory and field tests on self-healing substances and techniques of self-healing assessment were performed. Figure 10.1 illustrates a flow tree showing the various approaches to the advancement of cemented self-healing material. The diagram shows the differences between the natural and human processes. In this research, artificial techniques of bacteria induction and material supplementation were implemented to leverage the natural responses of bacteria (Noshi and Schubert 2018).

Evidence of participation of microorganisms in carbonate precipitation has led to the growth and implementation of bioprocess technology in the self-healing of building materials and has thus proved a solution to many problems that are affected by decay (Hosseini et al. 2011; Abo-El-Enein et al. 2013; Irwan et al. 2017; Irwan and Teddy 2017; Jagannathan 2018). The latest technology for the formation of a high-intensity cemented product is microbiologically induced calcium carbonate precipitation (MICP). Biocement serves as a means of mitigating CO_2 produced during cement production (Hosseini et al. 2011; Abo-El-Enein et al. 2013; Irwan et al. 2017; Irwan and Teddy 2017; Jagannathan 2018). Calcium carbonate is one of the most recognised minerals deposited in the phenomenon known as biocementation or calcite precipitation (MICP) caused in the microbiological domain. Recently, such deposits have appeared in order to protect and consolidate different construction materials as promising binders. Enhanced microbial concrete or mortar calcite precipitation has become a significant field of study in building materials. Calcium carbonate precipitation in the form of calcite can be enhanced by bacteria. Precipitation of calcium carbonate is a result of a prevalent microbial metabolic process that helps to produce the microbial precipitation of calcite. By manufacturing calcium carbonate, bacteria added into concrete have been shown to enhance the concrete properties. Several researchers worldwide have demonstrated this. Currently, biocement is an eco-friendly, sustainable energy-saving material which uses non-toxic bio-based processes, whereby microorganisms can be activated to perform a structural function. Key advantages include the following:

- Biocement manufacturing saves more energy and is eco-friendly to the environment.
- Harvesting this natural, untapped resource can lead to a completely novel approach to issues relating to geotechnical or environmental engineering.
- Ultra-high-performance concrete can be developed using biocement (Reddy et al. 2015; Achal and Mukherjee 2015; Veerappan and Chandru 2016).

Numerous researchers worked with certain improbable allies to find the best additive to carry out this complex technology, which appeared to be bacteria. For most microorganisms an alkaline environment is not ideal; interestingly the production of calcite ($CaCO_3$) has been shown with some alkaliphilic *Bacillus* species. These types of bacteria must, however, be applied manually to the crack formation, which means that a regular structural inspection is still necessary. The bacteria and their food supply (calcium lactate) were integrated into the concrete and studied by Jonkers et al. (2010). Activation of the microbes only occurred when water seeped into the cracks. This hydration mechanism activated the bacteria to start oxidising and forming calcium lactate. The insoluble $CaCO_3$ is then precipitated, interconnected and repaired on the surface of the crack (Knoben 2011).

As an advantage, the combination of elevated pH and CO_2 generated by the bacteria contributes to a further formation of carbonate. Bacterial spores can live without water or food for centuries and can still be effectively woken up. Ideally, its self-healing features should remain the same as spores within concrete. Initially, Jonkers et al. (2010) found that after only 4 weeks, the bioconcrete could no longer be self-repaired. The researchers solved this issue, before they mixed in the concrete, incorporating the bacteria and food in the clay capsules. Lab tests show that the use of bioconcrete can extend the structure's life by 50 years (Knoben 2011).

This provides a strong motivation for further investigation and subsequent implementation of MICP for biocement improvement. A discussion on the method of biotherapy follows including the main elements concerned and significant variables influencing the biocement for self-healing concrete. In addition, envisaged applications, prospective benefits and constraints of MICP for soil improvement are covered. Finally, the main challenges and future perspectives are briefly outlined.

10.2 Mechanism of Self-Healing Biocement

The application of biocement promotes the self-healing capability of the concrete when microcracks appear through precipitation of calcium carbonate and may potentially enhance construction material. The calcium carbonate precipitation is caused by metabolic pathways such as urea hydrolysis and the oxidation of organic acids. However, urea hydrolysis has the potential to produce high amount of calcium carbonate crystals in a shorter time compared to the other metabolic processes (Huang et al. 2016).

There are two important components which enable the self-healing process of the biocement through urea hydrolysis, which is the urease-producing bacteria itself and a mineral precursor such as calcium lactate or calcium chloride which will be converted into calcium carbonate. Even though cracks in concrete structures have the potential to heal autogenously with hydration, this is only when the width of crack is less than 0.3 mm (Xu et al. 2018). Hence, biocementation will be useful to handle microcracks with wider a width. The microcracks, if left unhealed, would even lead to the corrosion of reinforcement steel as there is high potential for ingress water to penetrate deep into the concrete structure. Figure 10.2 shows the self-healing mechanism of the biocementation process.

10.2.1 Microbiologically Induced Calcium Carbonate Precipitation (MICP)

Biomineralisation is a process involving organisms in mineral formation as a consequence of cellular activity that stimulates the required physico-chemical conditions for growth and formation (Ben Omar et al. 1997), whereas cementation is a process whereby calcium carbonate either forms or deposits on a surface where they tend to attach to the substrate creating a cement-like substance (Achal et al.

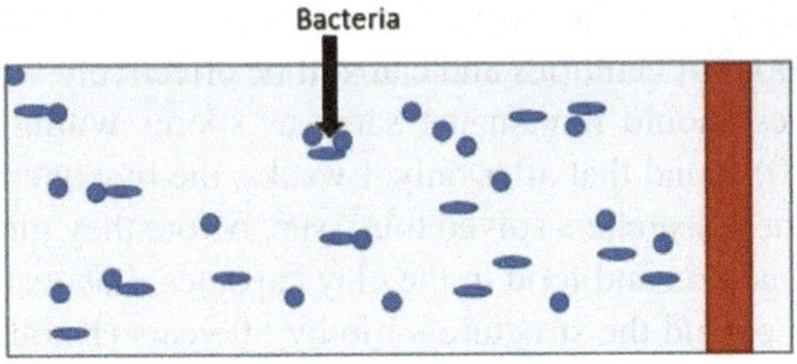

i) The biocement is mixed with spore forming alkalophilic bacteria to survive the extreme condition of the concrete, such as *Bacillus pasteurii.* This spore forming bacteria can stay inactive in the cement for up to 200years (Gavimath et al. 2012).

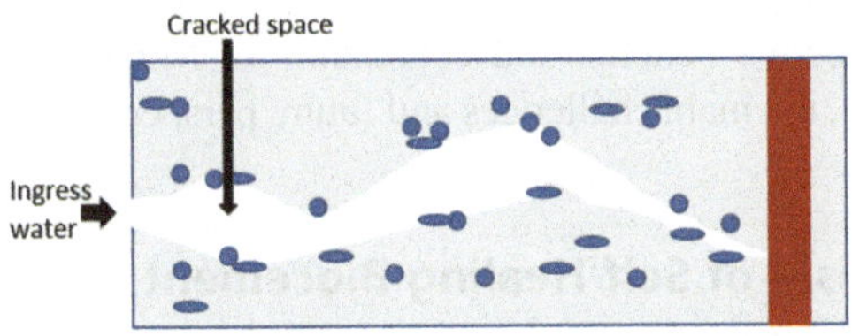

ii) Micro-cracking usually occurs in concrete structures due to thermal stress, creep, drying shrinkage, and general loading under serviceability conditions (Joseph et al. 2007). When a crack appears, it allows the passage of water inside the concrete structure, known as ingress water. The ingress water reaches the bacterial spores and allows them to germinate.

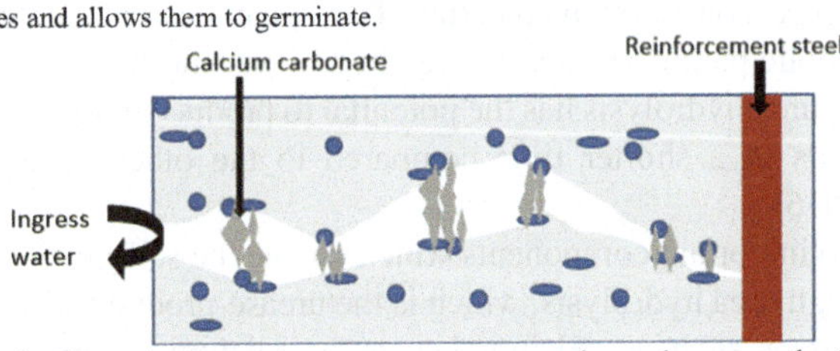

iii) Upon germination, the bacterium consumes urea through urease hydrolysis process and in turn releases ammonium and carbonate ions. When more carbonate ions accumulate in the bacterial environment, the calcium ions from the precursor such as calcium chloride react with the carbonate ions. This results in the precipitation of calcium carbonate or limestone at the cell surface. Overtime, more limestone accumulates, and it will seal the cracked spaces formed and thus preventing the ingress water to reach deep inside the concrete (Cuthbert et al. 2013). This process can be summarized in the equation below:

$$CO(NH_2)_2 + 2H_2O + Ca^{2+} \xrightarrow{\text{Urease from ureolytic bacteria}} 2NH_4^+ + CaCO_3 + 2Cl^-$$

Fig. 10.2 Self-healing mechanism of biocement. (Adapted from Jonkers 2018)

2015). Calcium carbonate can precipitate in three main polymorphs that are calcite, vaterite and aragonite (Wei et al. 2015). A concrete structure may possess high resistance to compression loads but possess a weakness in tension resulting in higher stress. Once the cracks are formed, the life span of concrete reduces. Various repair

Table 10.1 Different types of microbes potentially used for self-healing concrete

No	Microbes	References
1	*Bacillus pasteurii*	Vijay et al. (2017)
2	*Bacillus subtilis*	Vijay et al. (2017)
3	*Sporosarcina pasteurii*	Abo-El-Enein et al. (2013)
4	*Enterococcus faecalis*	Irwan et al. (2017)
5	*Bacillus sphaericus*	Jagannathan 2018
6	*Bacillus cereus* KLUVAA	Anitha et al. (2018)
7	*Yaniella* sp. VS8	Stabnikov (2016)
8	*Bacillus* sp. VS1	Stabnikov (2016)
9	*Bacillus lentus*	Parmar and Marjadi (2017)

techniques are available, but they are highly expensive and time-consuming processes.

Currently, bio-based solutions are a driver for smart materials such as self-healing concrete. MICP has been shown to adjust the mechanical and hydraulic properties of a porous material (De Jong et al. 2013). It was shown to fill the concrete cracks and bind the other materials such as sand and gravel in concrete. A crack size of more than 0.8 mm is difficult to repair; however with the use of bacteria, cracks can heal with the calcite precipitation (Luo et al. 2015). A study reported by Vashisht et al. (2018) shows an isolated *Lysinibacillus* sp. was able to enhance the comprehensive strength of concrete than the standard strain *Bacillus megaterium* MTCC 1684 after 28 days of curing. Recently, the urease activity has also been reported in fungi *Pestalotiopsis* sp. and *Myrothecium gramineum* (Li et al. 2015a, b) which was isolated from calcareous soil. Table 10.1 shows the list of different types of microbial potentially used for self-healing concrete.

MICP is an effective and environmentally friendly technology often used to solve concrete cracks and various environmental problems such as soil instability (Anbu et al. 2016). Therefore, in civil and geotechnical techniques, ureolytic bacteria are also being investigated for ground improvement (Omoregie et al. 2017). Specific bacteria living in the soil hydrolyse urea and release ammonium which causes the pH to rise that promotes the precipitation of calcium carbonate. The process is depicted in Eq. 10.1 (urea hydrolysis) and Eq. 10.2 (calcium carbonate precipitation) as explained previously by Cardoso et al. (2018):

$$(NH_2)_2CO + 2H_2O \rightarrow 2NH_4^+ + CO_3^{2-} \tag{10.1}$$

$$Ca^{2+} + CO_3^{2-} \rightarrow CaCO_3 \tag{10.2}$$

A study conducted by Omoregie et al. (2017) showed that the locally isolated *Sporosarcina pasteurii* from limestone cave samples of Sarawak were found to have high urease-producing activity that are capable of improving the strength of poorly graded soils. Achal et al. (2009) utilised a simplified MICP treatment technique for biocement production. The sand columns were injected with *S. pasteurii* (NCIM

2477) and cementation solution under gravimetric free-flow direction for the duration of 120 h. The findings show $CaCO_3$ contents precipitated in the sand columns and were mostly placed at the upper layer which led to the decrease of soil porosity and permeability.

10.2.2 Compressive Strength

Biocementation refers to a process in which particle-binding materials are generated through microbial activities which enhanced the shear strength of concrete (Jian and Ivanov 2009). It involved the deposition of calcium carbonate generated through a microorganism activity in the system rich in calcium ions through a bioprocess technology (Khanafari et al. 2011). Recent studies reported that the use of aerobic microorganisms such as *Pseudomonas aeruginosa* and *Bacillus pasteurii* as self-healing agents has shown 18% improvement in the compressive strength of cement mortar (Ramachandran et al. 2001). It is recommended to apply biocement-incorporated concrete instead of ordinary cement as the durability of concrete, which usually depends on its compressive strength, could be improved. This was proven by a study which showed an increment in the compressive strengths of mortars by 17% and 25% after 7 and 28 days, respectively, due to biocementation using *Shewanella* sp. (Ghosh et al. 2005).

The remediation of cracks in building materials may also contribute to developing rocks/bricks of different strengths and regain strength within a short time frame (Thawadi and Ruwisch 2012). The strength improvement was due to the biodeposition secreted by beneficial microorganisms within the pores of cement–sand matrix which indirectly act as self-healing agents. Earlier, Dick et al. (2006) performed biodeposition treatment in order to restore and protect degraded Euville limestone by using *Bacillus sphaericus*, which resulted in multiple layers of calcite deposited on the degraded limestone in 4 weeks. Basically, when cementitious materials were added with bacterial cells, they started growing after utilising nourishment from pores of cementitious materials and the surrounding medium. This mechanism known as microbiologically induced calcium carbonate precipitation (MICP) precipitates calcium carbonate, which occurred on the cell surface as well as within the cement mortar matrix (Rong and Qian 2012). Once the pores in the matrix start to be filled, the flow of the nutrients and oxygen to the bacterial cells is blocked. Eventually, this leads to cell death or turned into endospores and act as organic fibres (Ramachandran et al. 2001). This explained that besides MICP, the overall increase of strength also resulted from the presence of an adequate amount of organic substances in the matrix due to the microbial biomass. Overall, the increase in compressive strengths is mainly due to consolidation of the pores inside the cementitious materials with biomineralisation products coming from MICP as shown in Fig. 10.3.

A study by De Muynck et al. (2008) has shown that the durability of cementitious materials can be improved along with the deposition of carbonate by *Bacillus sphaericus* as surface treatment. A study also reported that by using *Bacillus*

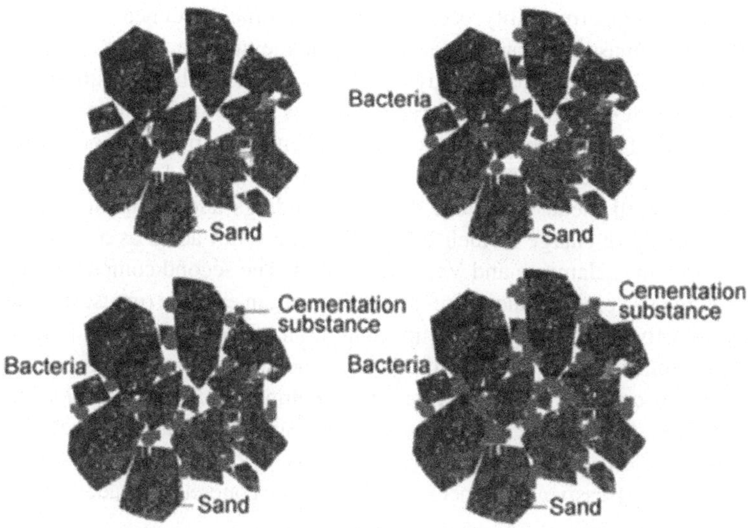

Fig. 10.3 Cementation mechanism of biocement. (**a**) Loose sand; (**b**) bacteria absorbed in the surface of loose sand; (**c**) cementation substance absorbed in the surface of loose sand; (**d**) loose sand to whole sand cemented by biocement. (Adapted from Rong and Qian (2012) (the image is directly from another article; check if allowed permission))

sp. CT-5, the compressive strength of cement mortar increased to 36%, which was able to enhance the durability of construction materials (Achal and Pan 2014).

10.2.3 Water Permeability

In a wet environment, the possibility of formation of crack on concrete structure is usually higher caused by the risk of intervention of aggressive substances. This is because of the penetration of aggressive substances responsible for concrete degradation. Generally, permeability depends on the pore network of cementitious materials which are quantified by variables such as porosity, size distribution, tortuosity, connectivity, specific surface and also microcracks (Phung et al. 2013). Most of the study found that MICP has exhibited an ability to considerably reduce water permeability of cementitious materials and other building materials by incorporating specific healing agents in the concrete matrix, indicated by a recent study which designed biological self-healing concrete by using *Bacillus sphaericus* (Tittelboom et al. 2010).

Another study, also utilising *Bacillus sphaericus*, reported that there is about 65–90% reduction in water absorption of the mortar specimens, due to the formation of a layer of calcite on the surface (De Muynck et al. 2008). A significant decrease in water permeability in concrete due to the biodeposition carried out by *B. sphaericus* promotes the crack repairing (Belie and Muynck 2009). Thus, it concluded that a decrease in water permeability occurring in the mortar specimens was due to the presence of biomass and carbonate crystals on the surface as well as inside the porous matrix. According to De Muynck et al. (2008), the biodeposition treatment had a unique pore-blocking characteristic which is made up from a two-component coating system. The first being that the bacteria themselves plug the pores and form a biofilm on the cementitious surface. As the bacteria inside the biofilm attract positively charged metal ions from their surroundings and act as nucleation sites due to the negative charge in their cell wall, this biofilm acted as a primer for the carbonate coating (Hammes and Verstraete 2002). The second component involved in this coating system was due to the increase in certain enzyme (urease and carbonic anhydrase) activities which lead to supersaturation of the liquid phase in relation to calcium carbonate. This results in the heterogeneous precipitation of calcium carbonate crystals on the biofilm which promotes complete healing of cracks occurring in the concrete. Overall, it clearly shows that resistance to water permeability plays an important role to enhance long-term durability of concrete as it determines the penetration of aggressive substances responsible for concrete degradation.

10.2.4 Microstructure

Biomineralisation can be described as the precipitation process of mineral materials influenced by living forms (Skinner and Jahren 2004). The presence of bacteria in a cement-based material generates the microbial activity which leads to the improvement of concrete properties. An understanding on the bacterial activity towards cement hydration and its chemical phase and microstructure mechanism will help in exploring the application of microorganism in concrete.

This microbiologically induced calcium carbonate precipitation (MICP) can generate carbonate ions. In a calcium-rich environment, the enzyme results from the process of urea hydrolysis (De Muynck et al. 2010) induced by precipitation of calcite (Ca) or calcium carbonate ($CaCO_3$). A recent study reported that the bacteria which had been proven to be able to precipitate CaCO3 are *Bacillus* or *Sporosarcina pasteurii* and *Bacillus sphaericus*. Another type is *Shewanella* species (Ghosh et al. 2009). Usually, microstructure tools are used to analyse the calcite formation through an MICP mechanism of calcium carbonate in a microbe cement-based material. Generally, scanning electron microscopy energy-dispersive X-ray (SEM-EDX) with the combination of other techniques is essential in visualising imaging, morphological information and mineralogical composition.

Recently, cement-based biomaterial has been developed to remediate the cracks in concrete structures. Previous studies have shown that the addition of specific microorganisms to cement-sand mortar or concrete deposit inorganic substances inside the pores of the matrices, can help to remediate cracks within the structures (Ghosh et al. 2005). Another study found out that with the addition of an anaerobic hot spring bacterium (closely related to *Shewanella* species) to the mortar/concrete could increase the compressive strength (25–30%) of the material with respect to control (Ghosh et al. 2006). The biologically induced cement-based materials thus also enhance better durability and crack-repairing performance compared to normal concrete materials (Ghosh et al. 2008). Thus, the application of effective microorganisms in a cracked concrete structure could remediate the cracks due to the formation of biodeposition filling up the micropore available in the concrete structure. Also the protein secreted by the bacterium leaches silica and helps in the formation of new silicate phases that fill the micropores. This protein increases the strength of mortar when it is added separately.

10.2.5 Chloride Ion Permeability

Chloride-induced corrosion due to chloride ion diffusion occurring in concrete is one of the main mechanisms of deterioration affecting the long-term performance of building structures. Steel reinforcement embedded in concrete is mainly protected against corrosion by passivation of the steel surface due to the high alkalinity of the concrete. However, the alkalinity of the concrete can be neutralised due to environmental effects, such as carbonation. Thus, the risk of corrosion increases if chlorides reach the steel reinforcement. In order to overcome this problem, the implementation of biocement plays a crucial role in sustaining the durability and performance of concrete. This is because MICP mechanism of biocement tends to block the pores which eventually prevent the progress of chloride ions. However, the presence of chloride ions usually causes corrosion of steel bars in a short time. Therefore, some researchers have evaluated alternatives to avoid this dilemma by considering calcium chloride as the source of calcium. Calcium nitrate was successfully used as an efficient calcium source by *S. pasteurii* (Qian et al. 2009).

Other than that, the presence of calcium chloride also resulted in higher production of ammonia, increasing the risk of reinforcement corrosion. Thus, Neville (1996) suggested utilising calcium lactate as its metabolic conversion does not result in production of massive amounts of ammonia. Basically, *Bacillus cohnii* was applied on the cracked concrete. The presence of calcium lactate in the concrete was utilised by this *Bacillus cohnii* which results in abundant amounts of 20–80-lm-sized $CaCO_3$ precipitates on crack surfaces which promotes the healing of the cracks (Jonkers et al. 2010). The previous study had compared the ability of calcium lactate and calcium glutamate to precipitate calcium carbonate induced by *B. cohnii* and found out that the $CaCO_3$ layer precipitated using calcium glutamate and produced larger thickness than that from calcium lactate (Xu et al. 2014). Another study recently reported that various calcium sources mainly known as calcium chloride, calcium oxide, calcium acetate and calcium nitrate influence calcite precipitation produced by *Bacillus* sp. Overall, they achieved the maximum yield of calcite precipitation with calcium chloride (Achal and Pan 2014).

De Muynck et al. (2008) reported that bacterial deposition of a layer of calcite on the surface of the mortar specimens resulted in a decrease of capillary water uptake, which lowers the water permeability towards gas. Generally, the presence of biomass contributed a huge impact on the overall decrease of the gas permeability in cement mortars and, thus, resulted in an increased resistance towards carbonation

(De Muynck et al. 2008). An improved resistance towards carbonation was due to the thickness of the carbonate layer (30–50 lm). Basically, this carbonate layer was generated from bacteria that have been added together with a calcium source or an increased concentration of calcium ions, resulting in the protective effect of the biodeposition treatment towards carbonation. De Muynck et al. (2008) reported that the biodeposition treatment based on MICP resulted in significantly lower chloride migration coefficients (10–40%) compared to the untreated specimens. Thus, this mechanism causes an increment in resistance towards the chloride migration of bacterially treated mortars which eventually increases the durability of the mortar.

10.3 Applications

10.3.1 Binder

In recent years, bacterial biomineralisation has been well established as a novel, environmentally friendly approach for the preservation of decay stones, especially those artworks and monuments that consist of carbonate minerals (Jroundi et al. 2017). In order to mitigate the effects of these detrimental decay processes, biocementation is being introduced as one of the useful approaches. Biocementation is a process that involves in producing a binding material based on MICP mechanism. It has been developed in the production of bacterial concrete, biological mortar and cracks in concrete remediation (Ariyanti et al. 2012). Generally, mortar meant a workable paste containing binder, aggregate and water that bind to the building material and fill the gaps. Biological mortar refers to a mixture of bacteria, limestone powder and nutrient medium that consists of calcium salt (Ariyanti et al. 2012).

The binder can be produced through MICP, and it confers several benefits, such as modifying the pore structure of the brick and dramatically reducing the ion permeability to water and chloride (Achal et al. 2009, 2010). MICP is useful as it can be applied in various building materials even after the construction. The mechanical properties can be modified by reduction of pore spaces and cementation between grains. For instance, MICP can be utilised to treat the surface of various building materials to seal the mouths of the pores to reduce the hydraulic permeability. Reducing hydraulic permeability is a pivotal factor to hinder environmental degradation and improve durability (Mukherjee and Achal 2014).

In the present study, microorganisms are added for precipitation of calcium carbonate on concrete and brick. The deposition diminishes permeability and corrosion in the substrate. Hence, improvement of hindering moisture ingress can be achieved (Mukherjee and Achal 2014). In this study, Portland cement is blended with bamboo leaf ash and pulverised burnt clay waste. Bamboo leaf ash and pulverised burnt clay waste are found to be suitable materials like pozzolan. The blended cement concrete had a low compressive strength compared to conventional concrete at early ages of hydration. However, the blended concrete had a higher strength at later ages (Temitopeab et al. 2015). In another similar study, agrowastes such as bamboo leaf and sugarcane leaf are used in the manufacture of low CO_2 binder.

There is interesting element in the agrowaste which is silicon involved in production of inorganic binders. Bamboo leaf and sugarcane leaf ashes have a high content of silica and other elements such as magnesium, potassium, sulphate and calcium. In addition, the percentage of chloride is reduced. This will be an advantage as chloride relates to the problem of corrosion in a steel-reinforced concrete (Roselló et al. 2015). As a binder, biocement can be utilised in making bio-sandstone with an improved degree of strength. It can be applied in strengthening the sand crust layer in desert or dust fixation in dirt tracks. In addition, it can help to stabilise sandy soil slopes, to construct aquaculture ponds and roads and to enhance the engineered and mechanical properties of soil (Li et al. 2015a, b). The most common material used for stabilisation of sand is cement, which works as a binder. However, alternatives to cement are being considered such as bio-based substrates that do not require high energy consumption and huge amount of raw material extraction and lead to land degradation. Thus, biocement emerged as a promising solution for replacement with cement grouting as it is an environmentally friendly substance (Naeimi and Haddad 2018).

Hydroxyapatite has been used extensively in biomedical applications as it possesses the great characteristics of biocompatibility. It has been found to be similar to the inorganic constituent phase of human and animal bones. For example, hydroxyapatite is utilised in fabrication of biocement for fixation of dental or orthopaedic implants. However, hydroxyapatite-based biocement is found to be fragile. The enhanced biocement should be a material that possesses good handling and plasticity index and is able to produce durable bond in this case. Therefore, gelatin, citric acid and malonic acid are combined together as a binder to strengthen the strength of hydroxyapatite-based biocements. As a result, the mechanical strength of hydroxyapatite-based biocement is enhanced with longer aging time in sultry environment, particularly when gelatin was engaged as a binding agent in biocement (López-Cuevas et al. 2018). As we know, the main component of an ordinary Portland cement is calcium silicate. It known as a poorly crystallised material which forms very small particles of submicron size. In contrast, calcium hydroxide is recognised as a well-crystallised material. The calcium hydroxide will precipitate any free space. However, it is mechanically weak in cement. Therefore, it utilises the reaction of phosphate compound with calcium hydroxide to produce hydroxyapatite. Recently, calcium silicate hydrate is added to hydroxyapatite to form a composite-like structure interspersed at a nanoscale level. Hence, the mechanical strength of biocement is being enhanced (Lu and Zhou 2007).

Mineral trioxide aggregate (MTA) emerged as a biocement for endodontic use. It is the highlight in clinical endodontic as it possesses great sealing ability, biocompatibility, tolerance to moisture, improved regeneration of peri-radicular tissues and antimicrobial effect (Song et al. 2006). It is a hydraulic binder that contains calcium silicate (C3S) with the addition of zinc helps to enhance the grind ability of the material which strengthens the connection with biocement's reactivity versus water.

Nevertheless, degree of whiteness and fitness of material could be greatly enhanced (Voicu et al. 2016).

Paper mill sludge ash is the waste from a paper recycling process. However, it consists of a lot of fibres which can be utilised in biocement production. In this study, biocement consists of paper mill sludge ash as a binder needs more water when involved in production of mortar (Hosseini et al. 2011). Other than that, application of chemical bonding technique is utilised to fabricate calcium phosphate bulk materials through a combination of α-tricalcium and amorphous calcium phosphates as the binder. The preparation of chemical bonding technique required water with binder paste, and then pastes will press into dense cylinders and hardening process happened after that. Eventually, hydrolytic conversion of α-tricalcium and amorphous calcium phosphates into hydroxyapatite crystals provides appropriate strength to press binder pastes (Kuznetsov et al. 2009).

10.3.2 Admixture

The microbiologically induced calcium carbonate precipitation (MICP) has been proven in various applications to enhance the strength and durability of building materials including soil (Achal et al. 2016). Affordable admixture can be produced through non-septic cultivation on low-cost organic materials or extraction of biopolymers from sewage sludge. These materials contain linear and branched polysaccharides and other proteins. The addition of xanthan, amylopectin and albumin to the cement showed a positive impact on enhancing the strength of concrete. Hence, admixture from sewage sludge is believed to play a highly important role in water retention and thickening effects (Stabnikov and Ivanov 2016). Sewage sludge is being utilised in the biocementation through blending its incinerated ash and co-combusting it with limestone with Portland cement (Hosseini et al. 2011).

Biopolymer is being utilised as an admixture to improve the properties of mortars, and wood straw, cactus juice and flour are being applied as composite biomaterials to improve the construction of clay (Vasanthabharathi 2017). In the present study, biocement as an admixture with fly ash (FA) was investigated to enhance the geotechnical properties of soils. Fly ash is the waste produced from coal in power generation that consists of silicate. Moreover, it is highly essential to utilise fly ash as it has been found to have an impact on soil texture, such as hydraulic improvement properties of soils (Li et al. 2018). Additionally, it brings a positive impact on soil amendment agent, agriculture, land reclamation and forestry (Jala and Goyal 2006; Lee et al. 2006). Therefore, in this research admixture of biocement with fly ash is proved to enhance the strength of the soil. However, as bacteria will produce calcite continuously, they support sustainable improvement of soil strength (Li et al. 2018).

Besides, rice husk and fly ash with bacteria have been proven to alleviate environment pollution as those wastes have been utilised in biocementation (Dhami et al. 2012). In addition, the bricks that are made by them with soil cement blocks will have a high reduction level of water absorption (Li et al. 2015a, b).

Bamboo leaf ash and pulverised burnt clay waste are blended with Portland cement. The result showed blended cement has high strength at later ages (Temitopeab et al. 2015). Various researches have been carried out by adding admixtures of cement-based materials to investigate the impact (Vieira et al. 2005). Chitosan has been chosen as an additive; it can help to enhance the engineered properties of biocements which included dental biocement and regeneration of the bone. Chitosan has been reported to act as a cohesion-enhancing agent in calcium phosphate cement for repairing the bone and serve as a cellulose derivative in cement polymer systems to increase the setting time. The result showed chitosan derivatives are able to act as an effective thickener, have water retention ability and reduce the workable life of the fresh paste (Lasheras-Zubiate et al. 2012).

10.3.3 Protective Layer

The microbiological process triggers slimes, calcium carbonate crystals and other minerals. Generally, biocement use on the surface of building materials decreases the permeation level as it will remove the calcium carbonate layer on the concrete surface. The water impermeability at the concrete without biocement layer is four times lower than the concrete with biocement layer. In a nutshell, biocement plays a highly pivotal role in reducing corrosion in reinforced concrete effectively by providing a protective layer of carbonate by calcite precipitation to reduce the pollutants transported inside the concrete and strengthen the durability of composite materials (Li et al. 2015a, b). Precipitation of crystals or slime served as cementing agent between sand grains and thus increase the mechnical strength of soil through reducing the pores inside the soil (Chu 2016).

Biocrust has been developed as a biocement to reduce the water conductivity of sand through bioclogging/biological clogging. This methodology utilises urease-producing bacteria to precipitate a calcium carbonate layer on top of the sand. This layer can be used for impenetrable for water storage and work as erosion control for beach or riverbank (Chu et al. 2015). However, it is reported to have a problem with brittleness of biocement. In order to overcome these limitations, biomimetic approach which involve the usage of combination of organic nanoparticle, and micromineral particles have been used. Additionally, hydroxyapatite and oxide of Si and Fe enhanced the hardness as a protective layer (Stabnikov and Ivanov 2016). In the previous studies, biocalcin coating and vaterite crystal from carbonatogenic bacteria and *Myxococcus xanthus*, respectively, help to reduce the water absorption of limestone and improve the consolidation of porous limestone (Reddy and Joshi 2018). In the early stages, biocement is applied on the surface of limestone monuments as a barrier layer, and it was proved to have a protective effect. Furthermore, extensive research is carried out for the deposition of a surface layer as a barrier to the construction materials which contribute to the consolidation and waterproof activity (Li et al. 2015a, b). In the present study, biocement used was a mixture of mineral salt, urea and bacterial suspension. The biocementation occurred and formed a water-impermeable and high strength crust layer on the surface of the sand, and this brings the advantages for the construction of acquaculture ponds in sands or dust fixation in the desert areas (Stabnikov et al. 2011).

10.3.4 Self-Healing

Cement-based materials are one of the most commonly used building materials. As time goes by, the lifetime structure of the cement is affected by environment conditions such as corrosion and thermal changes which create cracks in concrete (Morsali et al. 2019). The crack is detrimental to the durability of structures that resulted in shortened usage time and caused extra cost on the maintenance (Achal et al. 2016). The maintenance and repair cost for the cracks on the surface layer of concrete can cost millions (Ivanov and Stabnikov 2016). Cracking of concrete is directly linked with shrinkage and mechanical compressive and tensile forces of the material. As such, concrete structures are always reinforced with steel in order for the reinforcement to take over the external forces or stress.

Therefore, any pollutant substances or water can enter the concrete by passing through the crack, which disrupts the integrity of the structure (Morsali et al. 2019). Although the microcracks do not affect much on the strength properties of cement structure, they contribute to the material porosity and permeability. Ingress of chemical such as chlorides and acids may trigger degradation and premature corrosion of embedded steel reinforcement. Hence, the reliability and durability of cement structure are strongly affected (Jonkers 2011).

Hence, a bio-inspired approach based on biocement to have a self-healing effect has been developed to strengthen the durability of concrete structures. The self-healing effect is an autogenous technique, and it will be triggered after a defect on the materials occurred (Jonkers 2007). Therefore, the technology of self-healing concrete based on calcium carbonate precipitation has been developed. Bacterial concrete involves bacteria in the mortar and concrete to heal up the faults. Bacterial concrete emerged as a new generation of modified concrete which involved biocementation by microbiologically induced calcium carbonate precipitation, and it is being recommended for remediation of microcracks (Morsali et al. 2019). This precipitation on the crack surface helps in sealing and plugging the cracks in the concrete making deleterious materials and water much more difficult to enter the crack of composite materials (Jonkers 2007).

The principal mechanism of the autonomous self-healing effect is involving mineral-producing bacteria to repair cracks in the concrete. Then, the bacterial spores are embedded into the concrete along with the substrate in capsules of polylactic acid for protection. Once the crack is formed, water will pass through and break the dormancy of bacterial spores to produce active cells which produce calcium carbonate crystal to seal the cracks. Hence, the reliability and lifetime of the material can be prolonged (Jonkers 2018).

Calcium chloride and calcium lactate have been used as two components of a healing agent for the autonomous remediation of crack in the concrete (Jonkers 2007). In this process, the precipitation of calcium carbonate in the crack is directly produced from the conversion of calcium lactate in equimolar amounts of calcium carbonate and chemical reaction of metabolically produced carbon dioxide. The carbon dioxide produced will react with portlandite particle present in the crack interior. In addition, portlandite will react with carbon dioxide produced by bacteria. Thus, more calcium carbonate will be precipitated and contributed and enhanced self-healing effect on the crack (Jonkers 2011). In the present study, ureolytic bacteria are utilised for precipitation of calcium carbonate within porous media matrices. Microbiologically induced calcite precipitation offers low viscosity and low-pressure technique in protecting unwanted fluid or gas migration via fracture networks in fractured rocks and is found to be effective to seal the rock fracture and reduce the permeability up to 80% compared to control specimen (Bucci et al. 2016). In another study, polyurethane and silica gel are tested for enhancing self-healing concrete. Results showed calcium carbonate is precipitated in silica gel (25% by mass) and polyurethane (11% by mass) which contributes to high strength of regaining and lower water coefficient which indicates both are potential to be utilised as bacterial carrier for self-healing concrete cracks (Wang et al. 2012).

10.4 Challenges

The mechanism of microbial precipitation of calcium carbonate is yet another astonishing phenomenon which unlocks another unlimited potential of what microorganism are capable of. However, it is not easy to implement this mechanism effectively for the purpose of biocementation. There are some challenges that need to be overcome before the successful implementation of this green technology in civil engineering.

10.4.1 Ammonium Release

In biocement, the bacteria undergo a complex metabolic reaction and precipitate calcium carbonate and release ammonium as well. Calcium carbonate is the desired product which heals the cracks, but ammonium is not a desired product here. Ammonium is not environmentally friendly, and it also dissociates at pH above 8.2 and thus releases a toxic ammonia gas into the atmosphere (Constable et al. 2003). Hence, more research needs to be conducted in order to convert the ammonia to other products which are not hazardous to the environment.

10.4.2 Selection of a Suitable Bacteria Strain

Incorporating a living microorganism in extreme media such as cement is a challenging process indeed. The microorganism requires many factors such as suitable pH, temperature, nutrients and diffusion rate of metabolic products in order to survive. Only the microorganisms which could resist hard conditions such as spore forming and alkaliphile and pose urea metabolism activity can be selected for biocement purpose (Reddy et al. 2015). Besides this, another important attribute of the selected bacteria is that they need to be non-pathogenic and safe for humans because construction workers will get close contact with the biocement and humans will occupy the building made of biocement after the construction process.

10.4.3 Cost

The cost of biocement is not competitive compared to the conventional cement (Iezzi et al. 2019). The implementation of biocementation in civil engineering requires an industrial-scale production of the desired microorganism. The cost involved in the production of the bacteria will make the cost of biocement to be higher than the convention cement. So, the research to be developed includes a cost-effective biocement and should be carried until a solution is achieved.

10.5 Conclusion

This chapter summarises the capability of certain microbes in healing cracks in the concrete. Different types of microbes can be used for restoration of building materials. This unique environmentally friendly method of using microbes for self-healing concrete can be a good alternative for cost-effective biocement and prolong the life span of building concrete.

Acknowledgements The authors would like to express their sincere thanks for the support of the Ministry of Education (MoE) and UTM-RMC (Malaysia) through HICoE Grant No. RJ130000.7846.4 J262.

References

Abo-El-Enein SA, Ali AH, Talkhan FN, Abdel-Gawwad HA (2013) Application of microbial biocementation to improve the physico-mechanical properties of cement mortar. HBRC J 9:36–40

Achal V, Mukherjee A (2015) A review of microbial precipitation for sustainable construction. Constr Build Mater 93:1224–1235

Achal V, Pan X (2014) Influence of calcium sources on microbially induced calcium carbonate precipitation by *Bacillus sp*. CR2. Appl Biochem Biotechnol 174:307–317

Achal V, Mukherjee A, Basu PC, Reddy MS (2009) Strain improvement of *Sporosarcina pasteurii* for enhanced urease and calcite production. J Ind Microbiol Biotechnol 36:981–988

Achal V, Mukherjee A, Reddy MS (2010) Biocalcification by *Sporosarcina pasteurii* using corn steep liquor as the nutrient source. Ind Biotechnol 6:170–174

Achal V, Mukherjee A, Kumari D, Zhang Q (2015) Biomineralization for sustainable construction – a review of processes and applications. Earth-Sci Rev 148:1–17

Achal V, Mukherjee A, Zhang Q (2016) Unearthing ecological wisdom from natural habitats and its ramifications on development of biocement and sustainable cities. Landsc Urban Plan 155:61–68

Anbu P, Kang CH, Shin YJ, So JS (2016) Formations of calcium carbonate minerals by bacteria and its multiple applications. Springerplus 5:250–262

Anitha V, Abinaya K, Prakash S, Seshagiri Rao A, Vanavil B (2018) *Bacillus cereus* KLUVAA mediated biocement production using hard water and urea. Chem Biochem Eng Q 32:257–266

Ariyanti D, Handayani NA, Hadiyanto (2012) Feasibility of using microalgae for biocement production through biocementation. J Bioprocess Biotech 2:2

Belie ND, Muynck WD (2009) Crack repair in concrete using biodeposition. In: Proceedings of the international conference on concrete repair, rehabilitation and retrofitting, pp 291–292

Ben Omar N, Arias JM, Gonzalez-Munoz MT (1997) Extracellular bacterial mineralization within the context of geomicrobiology. Microbiologia 13:161–172

Bucci NA, Ghazanfari E, Lu H (2016) Microbially-induced calcite precipitation for sealing rock fractures. In Geo-Chicago, Chicago, pp 558–567

Cardoso R, Pires I, Duarte SO, Monteiro GA (2018) Effects of clay's chemical interactions on biocementation. Appl Clay Sci 156:96–103

Chu J (2016) Solutions to sustainability in construction: some examples. Procedia Eng 145:1127–1134

Chu J, Ivanov V, He J, Maeimi M, Wu S (2015) Use of biogeotechnologies for soil improvement. In: Ground improvement case histories. Elsevier, Amsterdam, pp 571–589

Constable M, Charlton M, Jensen F, Mcdonald K, Craig G, Taylor KW (2003) An ecological risk assessment of ammonia in the aquatic environment. Hum Ecol Risk Assess 9:527–548

De Jong JT, Soga K, Kavazanjian E, Burns S, Van Paassen, Bang SS et al (2013) Biogeochemical processes and geotechnical applications: progress, opportunities and challenges. Geotechnique 63:287–301

De Muynck W, Cox K, De Belie N, Verstraete W (2008) Bacterial carbonate precipitation as an alternative surface treatment for concrete. Constr Build Mater 22:875–885

De Muynck DW, Belie DN, Verstraete W (2010) Microbial carbonate precipitation in construction materials: a review. J Ecol Eng 36:118–136

Dhami NK, Reddy MS, Mukherjee A (2012) Improvement in strength properties of ash bricks by bacterial calcite. Ecol Eng 39:31–35

Dick J, De Windt W, De Graef B, Saveyn H, Van der Meeren P, De Belie N et al (2006) Bio-deposition of a calcium carbonate layer on degraded limestone by *Bacillus* species. Biodegradation 17:357–367

Flatt RJ, Roussel N, Cheeseman CR (2012) Concrete: an eco material that needs to be improved. J Eur Ceram Soc 32:2787–2798

Ghosh P, Mandal S, Chattopadhyay BD, Pal S (2005) Use of microorganism to improve the strength of cement mortar. Cem Concr Res 35:1980–1983

Ghosh P, Mandal S, Pal S, Bandyopadhyaya G, Chattopadhyay BD (2006) Development of bioconcrete material using an enrichment culture of novel thermophilic anaerobic bacteria. Indian J Exp Biol 44:336–339

Ghosh S, Chattopadhyay BD, Mandal S (2008) Use of hot spring bacteria for remediation of cracks and increment of durability of structures. Indian Concr J 1:11–16

Ghosh S, Biswas M, Chattopadhyay BD, Mandal S (2009) Microbial activity on the microstructure of bacteria modified mortar. J Cem Concr Compos 31:93–98

Hammes F, Verstraete W (2002) Key roles of pH and calcium metabolism in microbial carbonate precipitation. Rev Environ Sci Biotechnol 1:3–7

Hosseini MM, Shao Y, Whalen JK (2011) Biocement production from silicon-rich plant residues: perspectives and future potential in Canada. Biosyst Eng 110:351–362

Huang H, Ye G, Qian C, Schlangen E (2016) Self-healing in cementitious materials: materials, methods and service conditions. Mater Des 92:499–511

Iezzi B, Brady R, Sardag S, Eu B, Skerlos S (2019) Growing bricks: assessing biocement for lower embodied carbon structures. Procedia CIRP 80:470–475

Irwan JM, Teddy T (2017) An overview of bacterial concrete on concrete durability in aggressive environment. Pertanika J Sci Technol 25:259–264

Irwan JM, Anneza LH, Othman N, Alshalif AF, Zamer MM, Teddy T (2017) Mechanical properties of concrete with *Enterococcus faecalis* and calcium lactate. Procedia Eng 171:592–597

Ivanov V, Stabnikov V (2016) Construction biotechnology: biogeochemistry, microbiology and biotechnology of construction materials and processes. Springer, Singapore

Jagannathan P (2018) Studies on the mechanical properties of bacterial concrete with two bacterial species. Mater Today Proc 5:8875–8879

Jala S, Goyal D (2006) Fly ash as a soil ameliorant for improving crop production-a review. Bioresour Technol 97:1136–1147

Jian C, Ivanov V (2009) Biocement – a new sustainable and energy saving material for construction and waste treatment. Civil Eng Res 7:53–54

Jonkers HM (2007) Self-healing concrete: a biological approach. In: Self healing materials. Springer, Dordrecht, pp 195–204

Jonkers HM (2011) Bacteria-based self-healing concrete. Heron 56(1/2):1–12

Jonkers HM (2018) Bacteria-based self-healing concrete. Delft University of Technology, Delft

Jonkers HM, Thijssen A, Muyzer G, Copuroglu O, Schlangen E (2010) Application of bacteria as self-healing agent for the development of sustainable concrete. Ecol Eng 36:230–235

Jroundi F, Schiro M, Ruiz-Agudo E, Elert K, Martín-Sánchez I, González-Muñoz MT, Rodriguez-Navarro C (2017) Protection and consolidation of stone heritage by self-inoculation with indigenous carbonatogenic bacterial communities. Nat Commun 8:279

Khanafari A, Khams FN, Sepahy AA (2011) An investigation of biocement production from hard water. Middle-East J Sci Res 7:964–971

Knoben W (2011) Bacteria care for concrete. Mater Today 14:444

Kuznetsov A, Larionov D, Stepu A, Kuznetsov V, Putlayev V, Ivanov V (2009) Calcium phosphate scaffolds fabricated via chemical bonding technique from different precursors. Materialwissenschaft und Werkstofftechnik: Entwicklung, Fertigung, Prüfung, Eigenschaften und Anwendungen technischer Werkstoffe 40:277–284

Lasheras-Zubiate M, Navarro-Blasco I, Fernández JM, Alvarez JI (2012) Effect of the addition of chitosan ethers on the fresh state properties of cement mortars. Cem Concr Compos 34:964–973

Lee H, Ha HS, Lee CH, Lee YB, Kim PJ (2006) Fly ash effect on improving soil properties and rice productivity in Korean paddy soils. Bioresour Technol 97:1490–1497

Li Q, Csetenyi L, Paton G, Gadd GM (2015a) $CaCO_3$ and $SrCO_3$ bioprecipitation by fungi isolated from calcareous soil. Environ Microbiol 17:3082–3089

Li MM, Kawasaki S, Zhang QZ, Achal V (2015b) Novel microbial based low energy green building material production technology. J Adv Mater Res 1090:96–100

Li M, Fang C, Kawasaki S, Achal V (2018) Fly ash incorporated with biocement to improve strength of expansive soil. Sci Rep 8:25–65

López-Cuevas J, Vargas-Gutiérrez G, Rodríguez-Galicia JL, Rendón-Angeles JC (2018) Effect of some organic binders on the mechanical strength of hydroxyapatite-based biocements. MRS Adv 3:3729–3734

Lu D, Zhou S (2007) High strength biological cement composition and using the same. US Patent, US 2007/0098811 A1

Luhar S, Gourav S (2015) A review paper on self healing concrete. J Civil Eng Res 5:53–58

Luo M, Qian CX, Li RY (2015) Factors affecting crack repairing capacity of bacteria based self-healing concrete. Constr Build Mater 87:1–7

Morsali S, Isildar GY, Tahni A (2019) The application of bacteria as a main factor in self-healing concrete technology. J Build Pathol Rehabil 4:7

Muhammad NZ, Shafaghat A, Keyvanfar A, Majid MZA, Ghoshal SK, Yasouj SEM, Shirdar MR (2016) Tests and methods of evaluating the self-healing efficiency of concrete: a review. Constr Build Mater 112:1123–1132

Mukherjee A, Achal V (2014) A biological route for producing low energy binders. In: 23rd Australasian Conference on the Mechanics of Structures and Materials (ACMSM23), pp 9–12

Naeimi M, Haddad A (2018) Investigation on the environmental impact of soil improvement techniques: comparison of cement grouting and biocement. In: GeoShanghai international conference. Springer, pp 483–490

Neville AM (1996) Properties of concrete. Pearson higher education, 4th edn. Prentice Hall, Upper Saddle River

Noshi CI, Schubert JJ (2018) Self-healing biocement and its potential applications in cementing and sand-consolidation jobs: a review targeted at the oil and gas industry. In SPE liquids-rich basins Conference-North America. The Society of Petroleum Engineers

Omoregie AI, Khoshdelnezamiha G, Senian N, Ong DEL, Nissom PM (2017) Experimental optimisation of various cultural conditions on urease activity for isolated *Sporosarcina pasteurii* strains and evaluation of their biocement potentials. Ecol Eng 109:65–75

Pacheco-Torgal F, Labrincha JA (2013) Biotech cementitious materials: some aspects of an innovative approach for concrete with enhanced durability. Constr Build Mater 40:1136–1141

Parmar S, Marjadi D (2017) Bio cementation: a novel technique and approach towards sustainable material. World J Res Rev 4:36–41

Phung QT, Maes N, Schutter GD, Jacques D, Ye G (2013) Determination of water permeability of cementitious materials using a controlled constant flow method. Constr Build Mater 47:1488–1496

Qian CX, Wang JY, Wang RX, Cheng L (2009) Corrosion protection of cement-based building materials by surface deposition of $CaCO_3$ by *Bacillus pasteurii*. Mater Sci Eng C 29:1273–1280

Ramachandran SK, Ramakrishnan V, Bang SS (2001) Remediation of concrete using microorganisms. ACI Mater J 98:3–9

Reddy MS, Joshi S (2018) Carbon dioxide sequestration on biocement-based composites. In: Carbon dioxide sequestration in cementitious construction materials. Woodhead Publishing, Oxford, pp 225–243

Reddy S, Manjusha A, Arun Kumar M (2015) Bio cement–an eco-friendly construction material. Int J Curr Eng Technol 5:2277–4106

Rong H, Qian CX (2012) Characterization of microbe cementitious materials. Chin Sci Bull 57:1333–1338

Roselló J, Soriano L, Santamarina MP, Akasaki JL, Melges JLP, Payá J (2015) Microscopy characterization of silica-rich agrowastes to be used in cement binders: bamboo and sugarcane leaves. Microsc Microanal 21:1314–1326

Schlangen HEJG, Jonkers HM, Qian S, Garcia A (2010) Recent advances on self healing of concrete. In: FraMCos-7: proceedings of the 7th international conference on fracture mechanics of concrete and concrete structures, Jeju Island, Korea, 23–28 May 2010

Sikder A, Saha P (2019) Effect of bacteria on performance of concrete/mortar: a review. Int J Recent Technol Eng 7(6C2):12–17

Skinner HCW, Jahren AH (2004) Biomineralization. In: Biogeochemistry (Treatise on geochemistry, vol 8) (Schlesinger WH (ed)) 8:117–184

Song JS, Mante FK, Romanow WJ, Kim S (2006) Chemical analysis of powder and set forms of Portland cement, gray ProRoot MTA, white ProRoot MTA, and gray MTA-Agelus. Oral Surg Oral Med Oral Pathol Oral Radiol Endod 102:809–815

Stabnikov V (2016) Production of bioagent for calcium-based biocement. Int J Biotechnol Wellness Ind 5:60–69

Stabnikov V, Ivanov V (2016) Biotechnological production of biopolymers and admixtures for eco-efficient construction materials. In: Biopolymers and biotech admixtures for eco-efficient construction materials. Woodhead Publishing, Oxford, pp 37–56

Stabnikov V, Naeimi M, Ivanov V, Chu J (2011) Formation of water-impermeable crust on sand surface using biocement. Cem Concr Res 41:1143–1149

Temitopeab KJ, Olubunmiac OK, Olugbengaad A (2015) Strength of bamboo leaf ash and pulverized burnt clay waste blended cement concrete. IOSR J Mech Civil Eng 12:36–42

Thawadi SA, Ruwisch RC (2012) Calcium carbonate crystals formation by ureolytic bacteria isolated from Australian soil and sludge. J Adv Sci Eng Res 2:12–26

Tittelboom KV, Belie ND, Muynck WD, Verstraete W (2010) Use of bacteria to repair cracks in concrete. Cem Concr Res 40:157–166

Vasanthabharathi V (2017) Review on eco-friendly construction biotechnology. Int J Microbiol Res 8:25–28

Vashisht R, Attri S, Sharma D, Shukla A, Goel G (2018) Monitoring biocalcification potential of *Lysinibacillus sp*. isolated from alluvial soils for improved compressive strength of concrete. Microbiol Res 207:226–231

Veerappan K, Chandru P (2016) A brief review on biocement based concrete. Int J Eng Manage Res 6:142–144

Vieira MC, Klemm D, Einfeldt L, Albrecht G (2005) Dispersing agents for cement based on modified polysaccharides. Cem Concr Res 35:883–890

Vijay K, Murmu M, Deo SV (2017) Bacteria based self healing concrete-a review. Constr Build Mater 152:1008–1014

Voicu G, Badanoiu AI, Andronescu E (2016) Biocements with potential endodontic use. In: Nanobiomaterials in dentistry. William Andrew Publishing, Kidlington/Oxford, pp 79–106

Wang J, Van Tittelboom K, De Belie, Verstraete W (2012) Use of silica gel or polyurethane immobilized bacteria for self-healing concrete. Constr Build Mater 26:532–540

Wei S, Cui H, Jiang Z, Liu H, He H, Fang N (2015) Biomineralization processes of calcite induced by bacteria isolated from marine sediments. Braz J Microbiol 46:455–464

Xu J, Yao W, Jiang Z (2014) Non-ureolytic bacterial carbonate precipitation as a surface treatment strategy on cementitious materials. J Mater Civ Eng 26:983–991

Xu J, Wang X, Zuo J, Liu X (2018) Self-healing of concrete cracks by ceramsite-loaded microorganisms. Adv Mater Sci Eng 2018:5153041

Correction to: Microbiological Tools for Cultural Heritage Conservation

Amrita Kumari Panda, Rojita Mishra, and Satpal Singh Bisht

Correction to:
Chapter 8 in: A. N. Yadav et al. (eds.), *Microbial Biotechnology Approaches to Monuments of Cultural Heritage*, https://doi.org/10.1007/978-981-15-3401-0_8

The book was inadvertently published with an incorrect affiliation of the corresponding author Amrita Kumari Panda in Chapter 8 as Department of Microbiology, Gogate Jogalekar College, Ratnagiri, Maharashtra, India. The affiliation has now been corrected as Department of Biotechnology, Sant Gahira Guru University, Ambikapur, Chhattisgarh, India.

The updated online version of this chapter can be found at:
https://doi.org/10.1007/978-981-15-3401-0_8

A. N. Yadav et al. (eds.), *Microbial Biotechnology Approaches to Monuments of Cultural Heritage*, https://doi.org/10.1007/978-981-15-3401-0_11

The manufacturer's authorised representative in the EU is Springer
Nature Customer Service Centre GmbH, Europaplatz 3, 69115 Heidelberg,
Germany. If you have any concerns regarding our products, please
contact ProductSafety@springernature.com

Printed and bound by CPI Group (UK) Ltd, Croydon, CR0 4YY
30/09/2026
02225333-0002